Ralf Bruno Moura Lopes

# Biotechnological potential of Bacillus thuringiensis in the post-genomic era

Ralf Bruno Moura Lopes

# Biotechnological potential of Bacillus thuringiensis in the post-genomic era

2023

ScienciaScripts

This book is a translation from the original published under ISBN 978-620-5-50433-8.

Publisher:
Sciencia Scripts
is a trademark of
Dodo Books Indian Ocean Ltd. and OmniScriptum S.R.L publishing group

120 High Road, East Finchley, London, N2 9ED, United Kingdom
Str. Armeneasca 28/1, office 1, Chisinau MD-2012, Republic of Moldova, Europe
Printed at: see last page
ISBN: 978-620-5-73680-7

**Editorial Director:** Lidiane Vivaldini Olo
**Editorial Manager:** Aurora Jian
**Responsible Publisher:** Iulia Ciumac
**Editors:** Fabiola Bovo Mendonça, Mayra S. Hatakeyama Sato
**Editorial Assistant:** Anderson Tamakoshi
**Editorial Production Manager:** Ricardo de Gan Braga
**Review Manager:** Hélia de Jesus Gonsaga
**Review Coordinator:** Camila Christi Gazzani
**Reviewers:** Luciana Azevedo, Patrícia Cordeiro, Sueli Bossi
**Editorial Producer:** Roseli Said
**Iconography Supervisor:** Sílvio Kligin
**Iconography Coordinator:** Cristina Akisino
**Iconographic research:** Roberto Silva, Enio Rodrigo Lopes
**Text licensing:** Erica Brambila
**Arts Coordinator:** Aderson Oliveira
**Design:** Alexandre Santana de Paula
**Cover image:** www.ingimage.com
**Layout:** Elis Regina de Oliveira
**Assistant:** Jacqueline Ortolan
**Illustrations:** [sic] comunicação, Conceitograf, Cris Alencar, Ingeborg Asbach, Jurandir Ribeiro, Leo Teixeira, Ligia Duque, Luis Moura, Mário Yoshida, Osni de Oliveira, Paulo Cesar Pereira, Rodval Matias, Sonia Vaz, Studio Caparroz, Walter Caldeira
**Image Treatment:** Emerson de Lima
**Prototypes:** Magali Prado

# TABLE OF CONTENTS

# CHAPTER 1

## 1. INTRODUCTION

*Bacillus thuringiensis* is a Gram-positive, endospore-forming bacterium. The species is widely distributed, being found in a variety of environments, such as soil, insect interiors, leaf surfaces, and internal plant tissues. Most studies of *B. thuringiensis* have focused on direct effects against agriculturally relevant pest insects, although several new functions in plant protection against phytopathogens are being discovered. For example, the synthesis of zuithermicin A, an antibiotic that is mainly efficient against oomycetes and related phytopathogenic fungi. Another example is the production of N-acyl homoserine lactonase, which hydrolyzes N-acyl homoserine lactone (AHL), a signal molecule of the bacterial *quorum sensing* system, suppressing pathogen virulence.

Endophytic bacteria colonize the interior of the host plant without causing apparent damage. Successful endophytic colonization by *B. thuringiensis* has been reported in many plants, such as bean, corn, sugarcane, cotton, soybean, cabbage, and lentil. Costa et al. (2012) reported that 32.9% of cultivable endophytic bacterial isolates from the leaves of different cultivars of the common bean *(Phaseolus vulgaris)* belonged to the phylum Firmicutes, including different isolates of *B. thuringiensis*. However, much remains unknown about the association of *B. thuringiensis* with legumes. Studies on this association have shown that *B. thuringiensis* inoculated into the rhizosphere or seeds not only colonizes the interior of legumes, but also increases root nodulation and plant growth.

On the other hand, microbial diseases of the common bean are one of the main causes of reduced productivity of this important crop. The control of diseases caused by phytopathogens is generally done by a combination of crop management, pesticides and resistant cultivars. The application of pesticides is the main method of control, but its use is controversial because of the harmful effects on the environment and the selection of resistant pathogens. There are also microbial diseases for which there is still no proven efficient product. Thus, the use of microorganisms capable of protecting plants against phytopathogens has emerged as a more rational and safer alternative for crop management. In this context, endophytic bacteria can be useful for biological control of diseases, since they often occupy the same habitat as phytopathogenic microorganisms and cause no apparent damage to the host plant.

*B. thuringiensis* BAC3151 is an endophytic isolate from *P. vulgaris* that has demonstrated in *vitro* antimicrobial activity against phytopathogenic bacteria of bean, as well as potential for inhibition of *quorum sensing* Gram-negative bacteria. Indeed, genomic studies have revealed the genus *Bacillus* as a surprising potential source of antimicrobials, such as polyketides, non-ribosomal peptides as well as unusual antimicrobials and other secondary metabolites. These studies also allow research into events that change genome architecture, such as gene loss or duplication, rearrangements, and gene acquisition by horizontal transfer. In the present work, the genome of BAC3151 was sequenced to better investigate the genetic determinants potentially involved in the synthesis of antimicrobials that could be useful for the biocontrol of phytopathogens, as well as to analyze genomic differences of the isolate and other strains.

# CHAPTER 2

# 2. REVIEW OF LITERATURE

## 2.1. Endophytic Bacteria

Endophytic bacteria have been known for more than 120 years (HARDOIM et al., 2008) and correspond to microorganisms, cultivable or not, that colonize the interior of a host plant and cause no apparent damage to it (HALLMANN et al., 1997). Although nitrogen-fixing bacteria in the nodules of common bean *(Phaseolus. vulgaris)* have been well studied and characterized (MARTÍNEZ- ROMERO, 2003; MARTÍNEZ-ROMERO, 2009) and the endophytic bacterial community of the seeds and roots of this legume has been reported (LÓPEZ- LÓPEZ, et al., 2010), little is still known about the endophytic bacteria of the aerial tissues of *P. vulgaris*. Costa et al. (2012) described the endophytic bacteria culturable from the leaves of the Vermelhinho, Talismã and Ouro Negro cultivars of *P. vulgaris* and found that 36.7% of the isolates belonged to Proteobacteria, 32.9% to Firmicutes, 29.7% to Actinobacteria and 0.6% to Bacteroidetes. The results also showed a difference in the structure of the endophytic bacterial community according to the bean cultivar.

The entry of microorganisms into plant tissue can usually occur through stomata, lenticels, lesions and emergent surfaces of lateral roots and germinating rootlets (HUANG, 1986). However, the main entry ports for endophytic bacteria appear to be the lesions that occur naturally as a result of plant growth, in addition to root hairs and epidermal junctions (SPRENT and FARIA, 1988). Some bacteria also have active penetration into host tissues, producing cellulolytic and pectinolytic enzymes (HUREK et al., 1994).

Once inside the plant tissue, they can remain in a specific location, such as the root cortex, or colonize the plant systemically by transport through the conducting elements or the apoplast. The ability to disperse systemically has been demonstrated by endophytes such as *Erwinia* sp. on cotton (MISAGHI and DONNDELINGER, 1990), *Pseudomonas aureofaciens on* maize (LAMB et al., 1996), and *Bacillus thuringiensis* on legumes (TANUJA et al., 2013). Colonization of specific tissues by bacteria appears to be species specific. Generally, endophytic bacteria colonize the intercellular space (QUADT- HALLMANN and KLOEPPER, 1996), with few reports demonstrating intracellular colonization (FROMMEL et al., 1991; HUREK et al., 1994; QUADT- HALLMANN and KLOEPPER, 1996). Colonization of the vascular system has also been reported (GARDNER et al., 1982; LAMB et. al., 1996).

The colonization patterns of endophytic bacteria in plant tissues are strongly dependent on interacting biotic and abiotic factors (HALLMANN et al., 1997). One of the most important biotic factors is the presence of other plant-associated microorganisms, such as fungi, viruses and other bacteria (HALMANN et al., 1998). Since space and nutrients provided by the host are limiting factors, there are several types of interaction between microorganisms that have the same habitat, including antagonism, symbiosis, and mutualism. Internal plant tissues provide a more uniform and protective environment for microorganisms than the plant surface, where exposure to extreme environmental conditions such as temperature, osmotic potential, and ultraviolet radiation are limiting factors for survival (HALLMANN et al., 1997). An example of the influence of abiotic factors on the population of endophytic bacteria is what occurs on sites contaminated with heavy metals, petroleum, solvents, polycyclic aromatic hydrocarbons or other organic contaminants, where plants growing in these soils are able to recruit microorganisms that can degrade the pollutants. Soil characteristics such as pH, salinity and texture also affect endophytic bacteria indirectly, as they alter the bacterial community of the rhizosphere, the source of potential endophytic bacteria (DOTY, 2008; RYAN et al., 2008).

Regarding their life strategy, endophytic bacteria can be classified as obligate or facultative endophytes. The obligate endophytic bacteria are strictly dependent on the host plant for their growth and survival. The transmission of these bacteria occurs vertically or by means of a vector. Facultative endophytic microorganisms, on the other hand, can be characterized by alternating between the interior of the plant tissues and the environment. The endophytic microbial diversity found in plants can be explained mainly by the ability of several facultative endophytes to enter and persist inside the plant (ROSENBLUETH and MARTINEZ-ROMERO, 2006).

In addition, endophytic bacteria can provide many advantages for plants, including growth promotion (BARKA et al., 2002; TANUJA et al., 2013), induction of systemic defense mechanisms (MISHRA et al., 2006; BAKKER et al., 2007), reduction of disease symptoms caused by various pathogens (COOMBS et al, 2004; KLOPPER et al., 2004), protection against insects and nematodes (D'ALESSANDRO et al., 2014), production of anti-herbivory compounds (SCOTT, 2001; SULLIVAN et al., 2007), biological nitrogen fixation (MARTÍNEZ et al., 2003; JHA and KUMAR, 2007), and increased tolerance to environmental stresses (ANDREOLLI et al., 2013). Figure 1 presents the main types of metabolites and their roles, involved in the interaction between endophytic bacteria and plants.

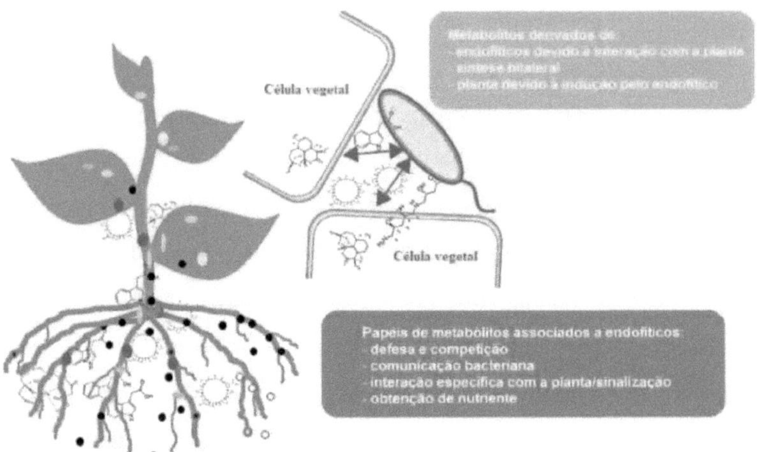


Figure 1. Schematic representation of different types of metabolites produced as a result of interactions between endophytic bacteria and plants. The metabolites produced by endophytic bacteria can exhibit important functions for the plant (BRADER et al., 2014).

## 2.2. *Bacillus thuringiensis*

### 2.2.1. General Features

*Bacillus thuringiensis* was first isolated from dead larvae of the silkworm *(Bombyx mori)* by bacteriologist Shigetane Ishiwata in Japan in 1901 while he was studying soto disease in this insect, and the species was initially named *Bacillus sotto* (ISHIWATA, 1901). However, in 1911, microbiologist Ernst Berliner isolated a related strain in larvae of the flour moth (*Anagasta kuehniella*) in the German state of Thuringia and subsequently named it *B. thuringiensis, a* name that has been retained (BERLINER, 1911; BERLINER, 1915).

*B. thuringiensis* is a Gram-positive, facultative anaerobic, mesophilic and chemoheterotrophic bacterium, growing mainly in the range 10 to 40 °C. It is shaped like a rod with dimensions of about 1.1 pm wide and 4.0 pm long, and is able to move around due to the presence of peritrichous flagella (depending on the serovar) (STAHLY et al., 1991). When under unfavorable environmental conditions, cell division is interrupted and the typical sporulation cycle of bacilli begins, generating an ellipsoidal spore located in the central or paracentral region of the cell. Simultaneously with the sporulation process, crystalline inclusions, characteristic of each *B. thuringiensis* variety and toxic to insects, are produced (FEITELSON et al., 1994; IBRAHIM et al., 2010). These crystalline inclusions, also called Cry proteins or ô-endotoxin, are released with the spore upon cell lysis (HANNAY and FITZ-JAMES, 1995). Moreover, they constitute an important characteristic to differentiate *B. thuringiensis* from similar species (FEDERICI, 1999).

### 2.2.2. Distribution in the environment

The species *B. thuringiensis* is widely distributed, being found in almost all parts of the world and associated with various environments, such as soil, water, stored grains (BRAVO et al., 1998), dead and live insects (SUZUKI et al., 2004) and plant surface and interior (BRAVO et al., 1998; MONNERAT et al., 2003; COSTA et al., 2012).

However, there is no certainty about the primary habitat of *B. thuringiensis* (MEADOWS, 1993). One possibility is that the bacterium is originally an entomopathogen, since it would be unlikely to use a large amount of energy during sporulation for crystal formation that does not have great adaptive value. Since the spore and crystal released by the cell tend to remain together in the environment, after ingestion by a susceptible larva and subsequent death of the insect, germination of the spore may then occur (THOMAS et al., 2000; VILAS-BÔAS et al., 2000 and SUZZUKI et al., 2004).

A second hypothesis is that *B. thuringiensis* is natural to the soil. The germination of spores in sterile soil is described in several studies and depends on moisture, nutrient availability, pH, and the soil type itself (AKIBA, 1986). However, subsequent work has shown that the spores of *B. thuringiensis* may often not germinate in non-sterile soil (VILAS- BÔAS et al., 2000) and that in this type of soil, the death of inoculated vegetative cells is significantly higher than in sterile soil (THOMAS et al., 2000; FERREIRA et al., 2003). According to Arantes et al. (2002), this lower viability may be due to a low capacity of *B. thuringiensis to* multiply in this environment, possibly because it is not an original member of the soil microbial communities, remaining mainly in the form of spores.

Another possibility less discussed, but presented by Smith and Couche (1991), is that the species is native to the phylloplane of trees. The authors isolated *B. thuringiensis* from several tree species, and in quantities large enough to be spread only by wind or rain. According to the authors, the soil is only a reservoir, while the primary habitat could be the tree phylloplane.

### 2.2.3. Use for biocontrol in agriculture

*B. thuringiensis* is a species that produces a variety of compounds released into the extracellular medium that are potentially useful for different types of biocontrol in agriculture. These compounds include antibiotics (FICKERS, 2012), chitinases (GOMAA, 2012) and other antifungals (SADFI et al., 2001), bacteriocins (CHERIF et al., 2003), lactonases (DONG et al., 2004), siderophores (BODE, 2009), and the characteristic o-endotoxin (Cry proteins), which constitutes the most studied and most widely employed product of *B. thuringiensis* in the world (POLANCZYK and ALVES, 2003).

The first study using *B. thuringiensis* for biocontrol in the field was conducted by Husz (1929), as part of an international program to control the

European corn caterpillar *(Ostrinia nubialis)*. Although at that time the cause of the entomopathogenic activity of the bacterium was not yet known, the results obtained were promising and from then on interest in the species grew. The first formulation based on *B. thuringiensis, Sporeine,* was produced in France in 1938. In the 1950s, large-scale production began with the launch of the commercial *bioinsecticide Thuricide,* followed by similar products from companies located in Russia, Germany, the United States and the former Czechoslovakia (DE MAAGD et al., 1999). Initially, the products were used only for the control of lepidopteran larvae. However, mainly since the 1970s, new subspecies of the bacterium have proven efficient against insects of the orders Diptera, Coleoptera (WEISER, 1986; EDWARDS et al., 1988), Hymenoptera, Hemiptera and Ortoptera (FEILTEISON, 1994). Toxicity has also been reported for nematode species, protozoa, mites and mollusks (FEILTEISON, 1994; WANG et al., 2013; LIU et al., 2014).

The efficacy and specificity of *B. thuringiensis* strains and their toxins in insect control favor the formulation of biopesticides based on this bacterium and more than a hundred formulations have been placed on the world market since the first product launched, accounting for more than 90% of total sales with bioinsecticides (POLANCZYK and ALVES, 2003). The direct application of crystals and spores of *B. thuringiensis* represents the main form of use of these biopesticides, which are harmless to the environment, beneficial insects and other animals, including humans (CÁRDENAS et al., 2001; BRAR et al., 2006).

In North America, products based on *B. thuringiensis* are widely used to control forest pests, especially *Lymantria dispar, Choristoneura fumiferana,* and *Choristoneura occidentalis, in* addition to insects harmful to agricultural crops. In Australia, these bioinsecticides are used against pests of crops such as cotton, fruit plants, ornamentals, and tobacco (GLARE and

O'CALLAGHAM, 2000; VAN FRANKENHUYZEN, 2000). They are also used against insect pests of major crops, forests, and vegetables on approximately one million hectares covering about thirty provinces in south-central China. In Egypt, *B. thuringiensis* is mainly used to control cotton pests (SALAMA and MORRIS, 1993). Cuba and Mexico lead the use of bioinsecticides based on this microorganism in Latin America, especially for the management of insects in cotton, banana, potato, citrus, vegetables, tobacco, corn and pasture crops. In Brazil, about thirty pests of agricultural importance and mosquito vectors of diseases are controlled *B. thuringiensis* (POLANCZYK and ALVES, 2003).

Despite the wide use, the cost of bioinsecticides based on *B. thuringiensis*, in most cases, is higher than that of chemical insecticides, the crystals cannot penetrate the plants and the persistence of most formulations in the field is low (GELERNTER and SCHWAB, 1993; VAN FRANKENHUYSEN, 1993; NAVON 2000). In this sense, research to minimize these limitations are being developed, mainly through the prospection of new *cry genes,* the production of plants expressing these genes, the development of new formulations and also the expansion of knowledge of the interactions of *B. thuringiensis* with other entomopathogens and chemical insecticides (NAVON, 2000).

## 2.3. Brief history of genomic science

With the advent of the DNA sequencing technology based on didesoxynucleotides developed by Sanger et al. (1977b), the progress of the first genome sequencing projects was made possible. This methodology made possible that, in the same year, the complete sequencing of the DNA of the bacteriophage 0X174 (SANGER et al., 1977a). Despite this initial advance, it was only after years of technological development using the methodology that the first genomes of cellular organisms, the bacteria *Haemophilus influenzae* and *Mycoplasma genitalium,* were completely sequenced (FLEISCHMANN et al., 1995; FRASER et al., 1995). Since then, the genomic science began to develop together with bioinformatics, completely changing the biological scenario and characterizing the most diverse organisms present in the tree of life.

Mainly due to the smaller genome size, the domains *Bacteria* and *Archaea* have the largest number of organisms with finalized genomes and ongoing genomic projects. There are a total of 39183 bacterial genomic projects currently in progress, with Proteobacteria, Firmicutes, and Actinobacteria accounting for 43.8%, 30.8%, and 14.5%, respectively, of these projects (https://gold.j gi-psf.org/distribution) (data as of December 2014) (Figure 2). The predominance of these phyla is mainly explained by the priority of sequencing the genome of organisms already previously studied and characterized by research groups around the world (KYRPIDES, 2009).

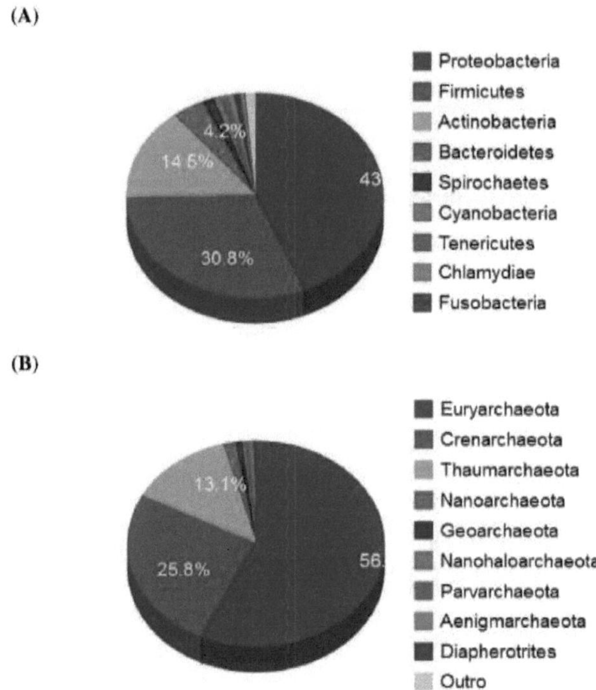



Figure 2. Phylogenetic distribution of (A) *Bacteria* and (B) *Archaea* genomic projects registered in the *Genomes Online Database* (GOLD) (https://gold.jgi-psf.org/distribution) as of December 2014.

Since the publication of the human genome *draft* (VENTER et al., 2001), the ability to generate genomic data at lower costs, faster and more efficiently has increased even more, especially due to the development of new DNA sequencing technologies, with the so-called next generation sequencers. There are several technologies available in the market aimed at large-scale DNA sequencing. These technologies differ mainly in terms of the mechanism employed in the sequencing process, the volume of data generated, the accuracy of bases, the size of fragments generated and the operational cost (MACLEAN et al., 2009).

In the first genome sequencing projects with Sanger technology, the fragments sequenced were small in number, up to 1000 bp in size, and the processing was done in periods of weeks or months, with strenuous laboratory work. However, at the beginning of this millennium, the automation of sequencing technologies was improved to the point where fully automated equipment was generating sequences in parallel 24 hours a day, and large genome sequencing centers around the world housed dozens of such sequencing machines. This, in turn, led to the need to create new genome assembly algorithms, using sequences typically 35 to 1000 bases long, with sequencing error rates ranging from 0.5 to 15%, and which may still contain complex artifacts such as repeats (MARDIS, 2008).

The growing advance in technology has also led to the emergence of several consortia and large-scale projects around the world. The Human Microbiome Project was started with the aim of characterizing the entire microbial community associated with different tissues of the human body and the expected results may contribute to a better understanding of human physiology and predisposition to disease (TURNBAUGH et al., 2007). The *TerraGenome* project, on the other hand, arose with the intention of sequencing the complete metagenome of different soil samples around the world (http://www.terragenome.org/). Finally, the largest and most ambitious current genome sequencing project, the *Earth Microbiome Project,* started in 2010, aims to characterize 200,000 samples from different environments on the planet. Through metagenomic, metatranscriptomic and amplicon sequencing approaches, the researchers foresee the reconstruction of approximately 500 thousand microbial genomes and the construction of a gene atlas that will contain information and metabolic models specific to each biome sampled (http://www.earthmicrobiome.org/).

## 2.4. Genome assembly

The assembly of a genome consists of a set of procedures in which one tries to organize a large number of short DNA sequences in a linear space with the objective of representing the complete DNA molecule(s) that compose the genome of the studied species (DOHM et al., 2007).

The enormous amount of DNA sequence data generated by next-generation sequencing (NGS) technologies along with the artifacts and errors inherent in each sequencing technology challenge projects to assemble complete genomes of different species. Less than a decade ago, only the *Newbler* assembler was available for this purpose, being intended for the assembly of fragments generated by sequences produced by the Roche 454 sequencer *(proprietary software)*. Introduced in mid-2007, the hybrid version of the MIRA assembler (CHEVREUX et al., 2004) was the first open-use assembler developed to assemble segments from the 454 sequencer and mixtures of segments from the 454 sequencer and the Sanger method. In late 2007, the SHARCGS assembler (DOHM et al., 2007) was developed to assemble short segments from Illumina technology and was quickly followed by a number of other *software programs*. Currently, there are several options of assemblers capable of using different fragment sizes, numerous file formats, and that are applied to genomes of different complexities (HUNT et al., 2014).

Genome assembly algorithms work by taking all the DNA segments at once, aligning them with each other, and trying to identify regions where two read segments overlap. These overlaps can be incorporated linearly into the assembly process. The shorter the sequences, the greater the amount of overlaps needed in order to perform the assembly. In addition, the genomic coverage, i.e. the number of times a certain region of the genome is covered by read segments, contributes to increase the accuracy in the considered region (EKBLOM and WOLF, 2014).

The use of reading segments with paired ends, that is, with overlapping of DNA sequence in its two extremities, facilitates the process of obtaining assembled sequences. When two or more of the sequences obtained from the sequencing process are paired (for having paired extremities), larger sequences called *contigs* are generated. *In turn,* the *contigs* can be paired, generating even larger sequences called *scaffolds.* Regions with no overlapping are usually found, characterizing *gaps* (BAKER, 2012). The determination of the sequence present in *gaps* can be obtained through the amplification specifically of these regions with subsequent resequencing, making it possible to achieve the final assembly of the genome (ALKAN et al., 2011).

NGS technologies easily generate two to three billion DNA read segments with 100 copies each, which can be used to assemble the genome of the species (BAKER, 2012). The assembly naturally represents a highly complex challenge due to the small size of the initial sequences. During genome assembly, read fragments are usually aligned to a genomic sequence known as a "reference" for the assembly (HUNT et al., 2014). In the absence of a reference genome, the read sequences must be used for a *de novo* assembly. The term *de novo* comes from Latin and means "from the beginning." It therefore refers to the methods used to determine the complete DNA sequence when no genomic sequence is available for use as a reference. The different *de* novo assembly strategies have advantages and disadvantages in speed and accuracy when compared to each other. *De novo* assembly is almost always complex and difficult, particularly when the genome is large and has sequences that repeat many times, causing assembly failures (ALKAN et al., 2011). The decision to use the *de novo* or reference-based assembly strategy, if the latter is available, is based on biological application, cost, effort to achieve the required accuracy, and assembly time considerations (EKBLOM and WOLF, 2014).

New advances in the assembly process are expected with the integration of databases and with the exploration of multiple sequencing strategies, always with the purpose of facing the challenge of assembling complex genomes. As mentioned, genomes that are large and marked by the abundance of repeated sequences are still a major challenge for the development of assembly algorithms from short DNA sequences. Usually, several assemblers are combined to circumvent this problem. Meanwhile, the precision and length of sequenced fragments have been gradually increasing (ZHANG et al., 2011). The challenge is to optimize the assembly algorithms to be able to process the vast amount of information from NGS sequencing that have their particularities for processing, both by volume and by the specific data types of each sequencer (CHAISSON and PEVZNER, 2008).

## 2.5. Genomic Annotation

### 2.5.1. Automatic annotation

The interpretation of raw genomic data involves the identification and annotation of several elements present in DNA, such as protein-coding sequences, tRNAs and rRNAs, pseudogenes and repetitive DNA (MÉDIGUE and MOSZER, 2007). The interest of the scientific community in genomic annotation includes clarifying the biological data obtained and placing it in a context for understanding biological processes (STEIN, 2001). Data interpretation can be performed at the nucleotide, amino acid and biological process levels.

When the annotation is at the nucleotide level, the main purpose is the prediction of probable genes or *open reading frames* (ORFs) (LUKASHIN and BORODOVSKY, 1998; STEIN, 2001). Programs that perform this task usually feature statistical calculations to recognize intrinsic features of the coding regions (STEIN, 2001). The use of orthologous genes from other organisms is also an important strategy for gene prediction (STEIN, 2001; MÉDIGUE and MOSZER, 2007).

After the prediction of genes, it is important to infer which products they encode and in which biological processes they will act (STEIN, 2001). Several programs are used in these annotation steps, not only to identify products and their functions, but also to standardize nomenclatures in a universal way (REEVES et al., 2009). Automatic annotation provides a final map composed of several characterized genetic elements that will have its content enhanced in manual curation. Table 1 presents the main programs used in automatic annotation and their respective functions.

Table 1. Computational tools widely used in the annotation of bacterial genomes.

| Use | Program | Reference |
|---|---|---|
| Gene prediction | FgenesB Glimmer | http://www.softberry.com/ Delcher et al., 1999 |
| tRNA prediction | ARAGORN tRNA-Scan-SE | Lowe and Eddy, 1997 Laslett et al., 2004 |
| rRNA prediction | RNAmmer Rfam | Lagesen et al., 2007 Griffiths-Jones et al., 2005 |
| Prediction of repetitive elements | *Repeat Scout* | Price et al., 2005 |
| Prediction of motifs, domains and protein families | InterProScan | Hunter et al., 2009 |

| Signal peptide prediction | SignalP | Bendtsen et al., 2004 |
| --- | --- | --- |
| Prediction of promoters | BPROM PromScan BDGP | http://www.softberry.com/ Studholme and Dixon, 2003 Reese, 2001 |
| Prediction of pseudogenes | Consed | Gordon et al.,1998 |

### 2.5.2. Manual curation

Manual curation is a crucial step for the finalization and validation of the automatic annotation of a given genomic sequence (MÉDIGUE and MOSZER, 2007). It is in this step that human evaluation (of the curator) is necessary to correct and legitimize the genomic information previously predicted (STEIN, 2001). As a result of manual curation, there is an increase in the accuracy of the data that will be made available to the scientific community (REEVES et al., 2009) and, consequently, the decrease of a subsequent propagation of errors arising from automatic annotation (GUNDOGDU et al., 2007).

The realization of manual curation occurs with similarity search analyses in databases containing well-characterized and conserved proteins and thus a more detailed description about the product of a particular gene or gene family is provided (OVERBEEK et al., 2005). In addition, at this time it is possible to characterize the metabolic pathways, domains and conserved protein motifs present in the genome as well as to use a uniform denomination for enzymes and other products (RUST et al., 2002).

Some groups working with prokaryotic genomics have added efforts to perform even the manual reannotation of genomes already deposited in databases (GUNDOGDU et al., 2007; LUO et al., 2009). These efforts are based on needs such as updating the information in several genomes that are outdated in databases, providing more detailed descriptions for their genes and respective proteins (GUNDOGDU et al., 2007), as well as correcting or reducing errors arising from automatic annotation, which may be incomplete or designate false descriptions (MADUPU et al., 2010), thus making the genomic information available more accurate.

It is also known that in the databases there are about 30% to 40% of proteins without known function or hypothetical proteins (HANSON et al., 2010). A flaw in these databases is that in many of them, as in the case of *GenBank,* the information does not undergo periodic updates, becoming obsolete and leading to error in new annotations (SALZBERG, 2007). Many of the proteins annotated as hypothetical already have a description available today as well as conserved domains that can help in the description of their functions and their updating has been performed in the process of manual curation, both unpublished and linked to the re-annotation (GUNDOGDU et al., 2007; LUO et al., 2009).

Indeed, other approaches make use of the results obtained post-curation, such as comparative genomics approaches. Genomic comparison between related species or between strains of a given microorganism can then be performed with greater certainty (RUST et al., 2002).

## 2.6. *B. thuringiensis* genomic projects

Currently, there are 42 *B. thuringiensis* genomes deposited at the *National Center for Biotechnology Information* (NCBI) (http://www.ncbi.nlm.nih.gov/ genome/browse/), of which 15 correspond to closed genomes and 27 to *draft* genomes (genomes not fully assembled) (data up to December 2014).

The deposited genomes for the species range from 4.93 Mb (for *B. thuringiensis* JM-Mgvxx-63) to 7.09 Mb (for *B. thuringiensis* T01-328), and in the closed genomes, 135 plasmid sequences in total were identified and made available. In *draft* genomes, possibly some *contigs* or *scaffolds* may correspond to plasmids, leading to an increase in this total value as other genomes are closed, although certain strains may not have plasmids. Strains with a larger genome also have a higher number of genes, with *B. thuringiensis* T01-328 having 7518 genes, while *B. thuringiensis* JM-Mgvxx-63 has 4931 genes. The number of protein-coding sequences for these two strains is 7403 and 4719, respectively. The G+C content among all sequenced genomes does not vary much and remains between 34.5 and 35.5%.

The interest in the sequenced strains of *B. thuringiensis* is mainly concentrated for use in biocontrol, due to characteristics such as insecticidal (GUAN et al., 2012; LIU et al., 2013; MURAWSKA et al., 2013; PALMA et al., 2014), nematicidal (LIU et al., 2014), molluscicidal (WANG et al., 2013) and chitinolytic (MARCON et al., 2014) activity. Other investigations on some of the strains include use for genetic engineering, due to high transformation efficiency (HE et al., 2010; JEONG et al., 2014), the study of biofilm formation (ZWICK et al., 2012) or even involvement in human pathology (a particular case) (HAN et al., 2006).

## 2.7. Genome mining and bioprospecting of *Bacillus* products

Recent advances in genome sequencing have highlighted the genus *Bacillus* as an unexpectedly promising source of bioactive compounds. In fact, for some of them, such as Bacillus *subtilis,* more than 4% of the genome has been found to be potentially involved in the synthesis of compounds such as non-ribosomal peptides, polyketides, and unusual antimicrobials

(KUNST et al., 1997; ARGUELLES-ARIAS et al., 2009; CHEN et al., 2009).

Non-ribosomal peptide synthetases (NRPSs) are large multimodular enzymes that synthesize non-ribosomal peptides from the successive binding of activated amino acids. In addition to core domains that catalyze the specific reaction to incorporate an amino acid, NRPSs have more specialized domains that perform modifications to these monomers, such as epimerization, methylation, oxidation, reduction, formylation, and heterocyclization. The specialized domains allow NRPSs to synthesize a large number of diverse structures with a variety of biological functions that could not be obtained by the ribosomal machinery (FELNAGLE et al., 2008). The structural diversity is responsible for the different biological activities, targets and mechanisms of action of non-ribosomal peptides, enabling the biotechnological exploitation of these peptides (FICKERS, 2012).

Since the characterization of the first non-ribosomal peptide, gramicidin produced by *Bacillus brevis,* in the early 1970s, many others have been isolated and their biosynthetic *clusters* characterized. With the emergence of powerful molecular and bioinformatics tools, the identification of even new metabolites through genome mining has become a reality (ANSARI et al., 2004; VAN LANEN and SHEN, 2006). For example, genome studies of *B. thuringiensis* led to the prediction of a novel NRPS consisting of seven modules that could be involved in the synthesis of a heptalipopeptide similar to kurstacin (ABDERRAHMANI et al., 2011). The discovery of new natural compounds with genome mining is an encouraging sign, suggesting that this methodology could lead to the isolation of new molecules of biotechnological interest.

In turn, polyketide synthases (PKSs) produce the core structure of polyketides from acyl-coenzyme A monomers (JENKE-KODAMA and DITTMANN, 2009). Type I PKSs are similar to NRPSs, in that different catalytic domains are found in a single polypeptide chain, whereas type II PKSs present the catalytic domains in mono- or bifunctional proteins (HERTWECK et al., 2007). In addition to the core domains involved in biosynthesis, additional domains of PKSs such as ketoreductase, dehydratase, methyltrasferase, enoylreductase, and oxidase are responsible for the chemical modification of the growing polyketide. Type II PKSs often exhibit a cyclase domain responsible for the formation of aromatic structures (MEURER et al., 1997). Similar to non-ribosomal peptides, due to the versatile assembly mechanism, polyketides exhibit remarkable diversity in terms of both structure and biological activity. For example, genes involved in the synthesis of dificidin, macrolatin, and bacillin have been characterized in *Bacillus amyloliquefaciens* (CHEN et al., 2007; ARGUELLES-ARIAS et al., 2009). While dificidin exhibits a broad spectrum of antibacterial activity (WILSON et al., 1987; ZIMMERMAN et al, 1987) and has been promising in antagonism to *Erwinia amylovara*, a phytopathogen that causes fire blight in various rosaceae, such as apple and pear (ZWEERINK and EDISON, 1987), macrolatin and bacillin present a smaller spectrum and have not been associated with biocontrol in agriculture, but have potential for use in medicine (GUSTAFSON et al., 1989; PATEL et al., 1995).

In addition to non-ribosomal peptides and polyketides, several *Bacillus* can also produce interesting antimicrobial peptides synthesized by the ribosomes, the bacteriocins. Strains of *B. thuringiensis,* for example, synthesize a variety of bacteriocins and bacteriocin-like inhibitory peptides. The antimicrobial peptides of *B. thuringiensis* are mainly turicins, tocicins and entomocins, such as turicin B439 from B. *thuringiensis* B439 (AHERN et al., 2003), turicin 17 from *B. thuringiensis* NEB17 (LEE et al., 2009b), turicin H from B. *thuringiensis* SF361 (LEE et al, 2009a), turicin S from *B. thuringiensis* sv. *entomocidus* HD198 (CHEHIMI et al., 2007), tocicin from B. *thuringiensis* sv. *tochigiensis* HD868 (PAIK et al, 1997), entomocin 9 from *B. thuringiensis* sv. entomocidus HD9 (CHERIF et al., 2003) and entomocin 110 from *B. thuringiensis* HD110 (CHERIF et al., 2008). *B. thuringiensis* sv. *tochigiensis* BGSC4Y1 and *B. thuringiensis* sv. *monterrey* BGSC4AJ1 can produce subtilosin (READ et al., 2009). Other antimicrobial peptides, such as morricin 269, kurstacin 287, keniacin 404, and tolvorticin 524 from Mexican strains have also been reported (BARBOZA-CORONA et al., 2007; DE LA FUENTE-SALCIDO et al., 2008).

Examples of bacteriocins with potential for biocontrol in agriculture include ericin S, produced by *B. subtilis* A1/3, which is active against *Clavibacter michiganensis,* the causal agent of tomato bacterial canker, and the peptide Bac 14, produced by *B. subtilis* 14B isolated from rhizosphere, active against *Agrobacterium tumefaciens* (ABRIOUEL et al., 2011). Other applications of bacteriocins in agriculture are also possible. Plant growth promotion by turicin 17 produced by *B. thuringiensis* NEB17 has been demonstrated (LEE et al., 2009 a,b). Peptides with antifungal activity can also be applied in the control of plant decay and in the postharvest control of fruits and vegetables, as in the case of the peptide produced by *B. amyloliquefaciens* RC-2, active against *Colletotrichum demantium* and other phytopathogenic fungi and bacteria (ABRIOUEL et al., 2011).

In Gram-negative bacteria, the population density-dependent control mechanism of gene expression, *quorum sensing,* involves signaling molecules, especially N-acyl homoserine lactones (AHLs) (FUQUA et al., 2001), and phytopathogenic bacteria can use this mechanism in regulating the production of virulence factors (VON BODMAN et al., 2003). However, *quorum sensing* can be inhibited by enzymes that degrade AHL, such as lactonases. Dong et al. (2004) showed that strains of *B. thuringiensis* producing N-acyl homoserine lactonase suppress quorum *sensing-dependent* virulence of the bacterial pathogen *Pectobacterium carotovorum* (*Erwinia carotovora*). Alternatively, plants genetically engineered to express *aiiA*, the *Bacillus* N-acyl homoserine lactonase gene, have been studied. Overexpression of this gene in tobacco plants, for example, significantly reduced the incidence of wildfire caused by *Pseudomonas syringae* pv. *tabaci* (QUIÑONES et al., 2005). These results show the potential to exploit antagonists that inhibit the bacterial cell communication mechanism for disease control and prevention.

In addition, plant growth and production can also be seriously affected by abiotic stresses. Among these stresses, iron deficiency constitutes the main factor that leads to reduced productivity, especially in calcareous soils, where the solubility of iron is extremely low (KOBAYASHI et al., 2005). One of the most used mechanisms by microorganisms to acquire iron is the production and excretion of siderophores, which are low molecular weight chelators that bind to $Fe^{3+}$ with high affinity and transport it into the cell (ZAWADZKA et al., 2009). Thus, one way in which plant-associated microorganisms that are producers of siderophores contribute to plant health is through the complexation of these compounds with iron, making it less available to phytopathogens (CHEN et al., 2009). It has also been shown that in experimental conditions of sterile soil, plants show symptoms of iron deficiency and have very low levels of the element in the root, which highlights the importance of microorganisms for the acquisition of iron for the plant (RADDADI et al., 2007). For *Bacillus,* the production of several siderophores was detected, mainly schizoquinene from *Bacillus megaterium* and others from *B. subtilis, Bacillus licheniformis* and the *B. cereus* group. Bacillibactin and petrobactin, for example, are produced by *B. cereus*

(PARK et al., 2005), *Bacillus anthracis* (CENDROWSKI et al., 2004) and *B. thuringiensis* (WILSON et al., 2006).

Another class of *Bacillus* products are the chitinases. Chitin is an insoluble polysaccharide formed by linear chains of N-acetylglycosamine residues joined by 0-1,4-glycosidic bonds. It is one of the most abundant biopolymers on the planet, second only to cellulose, and is an important component of the cell wall of fungi, as well as being found in the external skeleton of insects and also in algae, crustaceans, and internal structures of other invertebrates (BHATTACHRYA et al., 2007). Each year, a vast amount of chitin waste is released into the environment, which can generate environmental problems, and the decomposition of this polymer is important to maintain the carbon and nitrogen balance of the ecosystem (HAYES et al., 2008). Chitin catabolism involves the initial cleavage of the polymer by chitinases, producing oligosaccharides, and then subsequent hydrolysis into N-acetylglycosamine monomers by chitobiases (SUGINTA et al., 2000). These derivatives have a variety of agricultural, industrial, and medical applications, including use in cosmetics and antitumor activity (BANSODE and BAJEKAL, 2006).

Several *Bacillus* species, such as Bacillus *alvei, Bacillus lentus, Bacillus licheniformis, Bacillus cereus,* and *B. thuringiensis,* can produce chitinases during growth (ROJAS-AVELIZAPA et al., 1999; GOMAA, 2012). Chitinases are used as a potential biocontrol agent for many pathogenic fungi and thus can be useful for plant disease control (TANG et al., 2012), food and seed preservation (DE LA VEGA et al., 2006), as well as bioremediation (WANG and HUANG, 2001) and preparation of oligosaccharides and N-acetyl-D-glycosamine (WANG et al., 2006). Additionally, the potential role of chitinases in the biocontrol of insects, such as mosquitoes, has been suggested (HAYES et al., 2008). Chitinases can also be employed in the area of human health, as in the composition of ophthalmic preparations, and in other biotechnological areas (DAHIYA et al., 2006).

Thus, many *Bacillus* products have been discovered in recent years, and although great advances in their bioprospecting have occurred, new compounds of biotechnological interest are still expected to be found as well as the optimization of their production.

# CHAPTER 3

# 3. MATERIAL AND METHODS

## 3.1. Bacterial isolate and growth conditions

*Bacillus thuringiensis* BAC3151 belongs to the bacterioteca of the Laboratorio de Genética Molecular de Micro-organismos/BIOAGRO of the Universidade Federal de Viçosa (UFV). The isolate was previously obtained from *P. vulgaris* leaves (COSTA et al., 2012) and the grown in 10% TSA *(Tryptone Soy Agar)* medium (tryptone 1.5 g/l, soy peptone 0.5 g/l, NaCl 1.5 g/l, pH 7.3) at 28°C. The stock was kept at -80°C in cryotubes containing 10% TSA medium and 40% glycerol.

## 3.2. DNA extraction and genome sequencing

For genome sequencing, total DNA was isolated from the culture grown for 16 hours using the *Wizard genomic DNA purification kit* (Promega), according to the manufacturer's instructions for Gram-positive bacteria. The purified total DNA (25,278 ug and A $/A_{260280}$ = 2.05) was sequenced by Macrogen Inc. (Seoul, South Korea) using the HiSeq 2000 (100 bp *paired-end*) sequencing system (Illumina). A total of 21,512,378 high-quality *reads* and 2,172,750,178 high-quality bases were obtained, yielding a coverage of approximately 355X.

## 3.3. Data treatment and assembly

The raw data obtained from sequencing were processed using the Phred-Phrap-Consed package (RASKO et al., 2008). The minimum quality value used in the *base-calling* program was Q20. Then, assembly was performed using *SOAPdenovo* version 2.04 (LUO et al., 2012).

### 3.4. Annotating the genome

The annotation process involved the use of several algorithms in a multi-step procedure. Structural annotation was performed using the following *software:* FgenesB: gene predictor (http://www.softberry.org/);

RNAmmer: rRNA predictor (LAGESEN et al., 2007) and tRNAscan-SE: tRNA predictor (LOWE and EDDY, 1997). Functional annotation was performed by similarity analyses with BLAST (ALTSCHUL et al., 1990), using the non-redundant (NR) and conserved domain (CDD) bank, and analysis with InterProScan (JONES et al., 2014) *(e-value* <1e-6). In addition, manual annotation was performed with Artemis (RUTHERFORD et al., 2000).

### 3.5. Phylogenomic analysis

Phylogenomic analysis of *B. thuringiensis* was based on the *core genes* of BAC3151 and 27 other strains obtained from NCBI at gov/genomes/. In the process, the *core genes* were aligned with MUSCLE version 3.8.31 (EDGAR, 2004), the mismatched regions of the alignment were masked and then removed by GBLOCKS (CASTRESANA, 2000). Next, the remaining regions from all the alignments were concatenated to form a large sequence alignment, which was then used as input to PHYLIP v. 3.696 (FELSENSTEIN, 1989). The tree was constructed using the Neighbor-Joining method with 1000 replicates.

### 3.6. Sorting and orientation of *scaffolds*

The determination of the order and orientation of the *scaffolds of the B. thuringiensis* BAC3151 genome was done with Mauve *software* v. 2.3.1 by means of the *Mauve Contig Mover* algorithm (DARLING et al., 2010), using the sequence of *B. thuringiensis* MC28 as reference and the default parameters.

### 3.7. Gene sharing between strains

*Efficient Database framework for comparative Genome Analyses using BLAST score Ratios* (EDGAR) (BLOM et al., 2009) was used to analyze the pangenome and genomic *core of B. thuringiensis as* well as the *singletons of* BAC3151. In the process, the results of reciprocal BLAST between genome groups are considered. Protein-coding sequences (CDSs) from a genome that do not have a *score ratio value (*SRV) *(*observed score/'maximum *score*) higher than the cutoff in any of the other genomes analyzed are considered *singletons.* The SRV of the cutoff point (0.33) was generated during the creation of the genome bank used in the analysis. In the *core* genome calculation, only the CDSs with the best reciprocal BLAST *hit* in all the genomes used are taken into account.

### 3.8. *In silico* gene *cluster* analysis of secondary metabolites and other putative bioactive products

To analyze gene *clusters of* secondary metabolites and other products of BAC3151, antiSMASH (MEDEMA et al., 2011) and BAGEL3 (VAN HEEL et al., 2013) *software* were used. Insecticide genes were searched with the BtToxin_scanner (YE et al., 2012). All programs were used with default parameters.

### 3.9. Nucleotide sequence accession number

The *Whole Genome Shotgun* project has been deposited in DDBJ/EMBL /GenBank with accession number LDKD00000000. The version described in this book is LDKD01000000.

# CHAPTER 4

# 4. RESULTS AND DISCUSSION

## 4.1. General characteristics of the *B. thuringiensis* BAC3151 genome *draft*

The *draft* genomic sequence of *B. thuringiensis* BAC3151 has 5,743,871 bp, distributed in 42 *scaffolds* (with an average size of 136,758 bp and N50 of 552.900 bp), and a total G+C content of 34.9% (A=1,867,003 bases [32.5%]; T=1,871,097 [32.57%]; C=1,008,063 [17.55%]; G=996,602 [17.35%]; N=1,106 [0.01%]). Table 2 shows the result of processing the *reads up* to the assembly of the *scaffolds*.

Table 2. Summary of processing data obtained from sequencing and assembly.

| | |
|---|---|
| Initial no. of *reads* | 24.391.936 |
| N° of filtered *reads* (>Q20) | 21.512.378 |
| No. of *contigs* | 112,0 |
| No. of *scaffolds* | 42,0 |

A total of 5562 protein-coding sequences (CDSs) were predicted, and the CDSs comprised 4,604,274 bp, resulting in a protein-coding percentage of approximately 80.2%. The G+C content of the CDSs was slightly higher than the overall and corresponded to 35.87% (A=1,596,789 bases [34.68%]; T=1,355,335 [29.43%]; C=689,264 [14.97%]; G=962,377 [20.9%]; N=509 [0.01%]). The gene density was 0.968 CDS/kb, with the average size of the

CDSs equal to 827 bp. The *draft* genome also had 53 tRNA and 4 rRNA coding genes. The main characteristics of this and the other *B. thuringiensis* genomes used in this work are presented in Table 3.

    *B. thuringiensis* strains have been shown to have similar G+C content, coding percentage and average length of the CDSs. However, significant differences are observed in genome size and also in the composition of strain-specific genes. Mobile genetic elements such as plasmids, transposons, phages, integrons and genomic islands have an important contribution to these differences (FROST et al., 2005; FANG et al., 2011; HE et al., 2011). In addition, 86 proteins from the predicted proteome of *B. thuringiensis* BAC315 did not generate significant alignment with the banks used *(e-value* <1e-6) and may be encoded by genes not yet characterized. This number is higher than that of other *B. thuringiensis* strains with available genome.

Table 3. General properties of *Bacillus thuringiensis* genomes used in this study.

| Organism | *Status* | Size (Mb) | Access Number | Reference |
|---|---|---|---|---|
| *B. thuringiensis* BAC3151 | *Draft* | 5,74 | LDK00000000.1 | This study |
| *B. thuringiensis* MC28 | Complete | 6,69 | CP003687.1, CP003688.1 CP003693.1, CP003690.1 CP003689.1, CP003692.1 CP003691.1, CP003694.1 | Guan et al. (2012) |
| *B. thuringiensis* Bt407 | Complete | 6,13 | CP003889.1, CP003896.1 CP003893.1, CP003897.1 CP003890.1, CP003894.1 CP003891.1, CP003898.1 CP003895.1, CP003892.1 | Sheppard et al. (2013) |

| *B. thuringiensis* BMB171 | Complete | 5,64 | CP001903.1, CP001904.1 | He et al. (2010) |
|---|---|---|---|---|
| *B. thuringiensis* HD-789 | Complete | 6,33 | CP003763.1, CP003769.1<br>CP003764.1, CP003766.1<br>CP003765.1, CP003768.1<br>CP003767.1 | Doggett et al. (2013) |
| *B. thuringiensis* YBT-1518 | Complete | 6,67 | CP005935.1, CP005937.1<br>CP005939.1, CP002486.1<br>CP005938.1, CP005940.1<br>CP005936.1 | - |
| *B. thuringiensis* sv. *thuringiensis* str. IS5056 | Complete | 6,77 | CP004123.1, CP004126.1<br>CP004127.1, CP004131.1<br>CP004128.1, CP004136.1<br>CP004137.1, CP004132.1<br>CP004135.1, CP004130.1<br>CP004134.1, CP004129.1<br>CP004124.1, CP004125.1<br>CP004133.1 | Murawska et al. (2013) |

Table 3. Continued.

| *B. thuringiensis* sv. *kurstaki* str. HD73 | Complete | 5,91 | CP004069.1, CP004071.1<br>CP004072.1, CP004073.1<br>CP004070.1, CP004075.1<br>CP004076.1, CP004074.1 | Liu et al. (2013) |
|---|---|---|---|---|
| *B. thuringiensis* sv. *konkukian* str. 97-27 | Complete | 5,31 | AE017355.1, CP000047.1 | Han et al. (2006) |
| *B. thuringiensis* str. Al Hakam | Complete | 5,31 | CP000485.1, CP000486.1 | Challacombe et al. (2007) |

| *B. thuringiensis* sv. finitimus YBT-020 | Complete | 5,68 | CP002508.1, CP002509.1, CP002510.1 | Zhu et al. (2011) |
|---|---|---|---|---|
| *B. thuringiensis* sv. chinensis CT-43 | Complete | 6,15 | CP001907.1, CP001913.1 CP001908.1, CP001917.1 CP001909.1, CP001915.1 CP001912.1, CP001910.1 CP001914.1, CP001916.1 CP001911.1 | He et al. (2011) |

## 4.2. Phylogenomic Analysis

Conventional markers for phylogenetic analysis, such as 16S rDNA, are essentially identical for *B. thuringiensis* strains (as well as others of the *Bacillus cereus* group) (BAVYKIN, et al., 2004) and the relationship between them using *housekeeping* genes is often inconsistent (HELGASON et al., 2004). Therefore, phylogenomic analysis was performed using 1276 *core genes* from the strains (Figure 3).

BAC3151 was grouped together with *B. thuringiensis* MC28 isolated from soil and which is highly toxic to lepidoptera and diptera (TAN et al., 2009; TAN et al., 2010). This result is consistent with the fact that soil is a potential source of endophytic bacteria (DOTY, 2008; RYAN et al., 2008). Other strains isolated from different environments were also grouped together. For example, strains obtained from soil, such as YBT-1518 and IS5056, clustered with strains isolated from insects (ATCC 10792, T01001 [ZWICK et al., 2012] and Bt407).

Previous studies based on *multilocus sequence typing* (MLST) (PRIEST et al., 2004), BOX-PCR *fingerprinting* (KIM et al., 2002), genome mapping (CARLSON et al., 1996), pulsed field gel electrophoresis (PFGE) (CARLSON et al., 1994), *multilocus* enzyme electrophoresis (MLEE)

(HELGASON et. al, 2000), variable tandem repeat number mapping (VNTR), amplified fragment length polymorphism (AFLP) (TICKNOR et al., 2001), and mass spectrometry-based proteomic techniques integrated with statistical methods (DWORZANSKI et al., 2010), revealed the extensive similarity among *B. thuringiensis* strains as well as among other *B. cereus* group strains. The clustering of functionally distinct strains (as insect pathogen and endophyte with antimicrobial activity) is in agreement with previous study of the phylogenetic relationship between *B. thuringiensis strains* based on the complete genome (ZWICK et al. 2012).

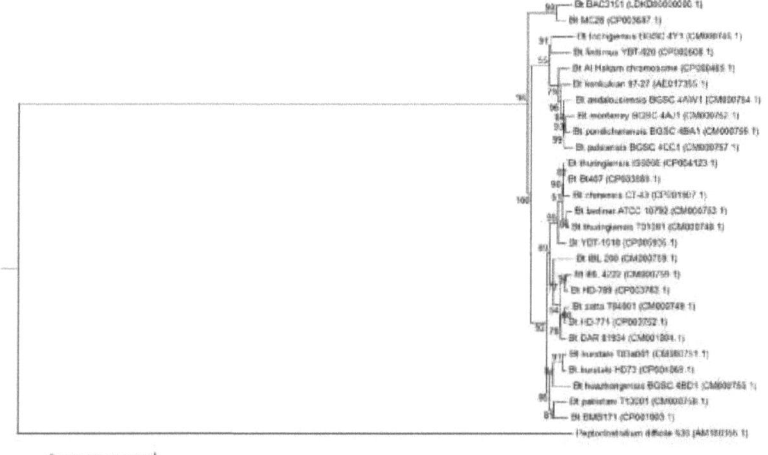

Figure 3. Phylogenomic analysis of *B. thuringiensis* based on 1276 *core* genes of the used strains. The tree was constructed with PHYLIP (FELSENSTEIN, 1989), using the NeighborJoining method with 1000 replicates. *Peptoclostridium difficile* 630 was used as an outgroup. *Bootstrap* values are shown at the nodes. The scale bar shows five nucleotide substitutions per 100 nucleotides. The accession number of the sequences used is given in parentheses.

## 4.3. Sorting and orientation of *scaffolds*

*B. thuringiensis* MC28, which was the closest strain to BAC3151 (item 4.2), was used as a reference to determine the order and orientation of the *scaffolds* of our isolate. Table 4 shows the final position of the *scaffolds* (numbered in pseudocoordinates) as well as the *forward* or *reverse* orientation. Despite the accuracy, some caution is needed in interpreting the results of the *scaffold* arrangements, since many true genomic rearrangements can occur in regions of repeated sequences, making sequence alignment difficult. Rissman et al. (2009) reported an accuracy between 90.4% and 99.4% of the ordering and orientation of sequences from different genome *drafts* using the *Mauve Contig Mover*.

Determining the order and orientation of the *scaffolds* facilitates the closure of *draft* genomes and comparative analyses. Using a reference genome more similar to the *draft for* this purpose produces more accurate ordering and greater coverage of sequences, and establishes the most likely orientation of the sequences.

Table 4. Order and final orientation of all *scaffolds of* the *B. thuringiensis* BAC3151 genomic *draft*.

| Name | Guidance | Starting position | Final position |
|---|---|---|---|
| scaffold9 | *reverse* | 1,0 | 287073 |
| scaffold4 | *forward* | 287074 | 351153 |
| scaffold26[a] | *forward* | 351154 | 1095058 |
| scaffold24 | *reverse* | 1095059 | 1647958 |
| scaffold16 | *forward* | 1647959 | 2063090 |
| scaffold6[a] | *reverse* | 2063091 | 2124415 |
| scaffold32 | *forward* | 2124416 | 2225266 |
| scaffold31 | *forward* | 2225267 | 2240649 |
| scaffold3 | *forward* | 2240650 | 2891218 |
| scaffold19 | *forward* | 2891219 | 3022029 |
| scaffold20[a] | *forward* | 3022030 | 3066338 |
| scaffold23[a] | *reverse* | 3066339 | 3080235 |
| scaffold5 | *reverse* | 3080236 | 4082243 |
| scaffold13[a] | *forward* | 4082244 | 4506425 |
| scaffold25 | *forward* | 4506426 | 4650005 |
| C1123[a] | *reverse* | 4650006 | 4664851 |
| C1173[a] | *forward* | 4664852 | 4713332 |

Table 4. Continued.

| Name | Guidance | Starting position | Final position |
|---|---|---|---|
| scaffold27 | *reverse* | 4713333 | 4771614 |

| scaffold2 | forward | 4771615 | 4837291 |
|---|---|---|---|
| scaffold18 | forward | 4837292 | 4858510 |
| scaffold1[a] | forward | 4858511 | 4887195 |
| C1055 | reverse | 4887196 | 4889722 |
| C1051 | reverse | 4889723 | 4892122 |
| C1083[a] | reverse | 4892123 | 4897873 |
| C1105[a] | reverse | 4897874 | 4905973 |
| scaffold28 | reverse | 4905974 | 5043094 |
| scaffold7 | reverse | 5043095 | 5125671 |
| scaffold21 | forward | 5125672 | 5334211 |
| scaffold12 | reverse | 5334212 | 5523281 |
| scaffold8 | forward | 5523282 | 5526453 |
| scaffold10 | forward | 5526454 | 5554308 |
| scaffold11 | forward | 5554309 | 5596095 |
| scaffold17 | forward | 5596096 | 5604454 |
| scaffold22 | forward | 5604455 | 5611028 |
| scaffold29 | forward | 5611029 | 5663233 |
| scaffold30 | forward | 5663234 | 5703273 |
| C1031 | forward | 5703274 | 5704386 |
| C1035 | forward | 5704387 | 5705583 |
| C1037 | forward | 5705584 | 5706865 |
| C1079 | forward | 5706866 | 5712279 |
| C1127 | forward | 5712280 | 5727434 |
| C1131 | forward | 5727435 | 5743871 |

· *Scaffolds* with alternative location possible.

## 4.3. Gene sharing between strains

The pangenome, *core* genome and the average number of *singletons* of *B. thuringiensis* were calculated based on the *draft genome of* BAC3151 and the 11 complete genomes described in Table 3. The pangenome of the analyzed strains was estimated to have 14755 CDSs, featuring 3413 *core CDSs,* which corresponds to approximately 49.3% to 71.1% of the CDSs in the genomes. The average number of *singletons* was 326 CDSs.

As more complete genome sequences of a species become available, the size of the pangenome of that species will generally increase due to an increase in the number of accessory genes. Based on mathematical models, it is predicted that new genes will be discovered within the pangenome of many bacterial species even though hundreds or possibly thousands of complete genomic sequences have been characterized (MEDINI, 2005). Figure 4A shows the increase in the number of CDSs in the pangenome of *B. thuringiensis* with increasing number of genomes analyzed. It can be seen that *B. thuringiensis* has an "open" pangenome. This observation is in agreement with the hypothesis that an "open" pangenome is the norm for the genus *Bacillus,* although *Bacillus anthracis* is the exception (FANG et al., 2011).

The reasons why exclusively the 12 genomes rather than all *B. thuringiensis* genomes available at NCBI were used for the calculations are twofold. First, the pangenome of the species is open. Second, for the *core genome* (Figure 4B) and *singleton* (Figure 4C) analyses, the curve remained about the same with the addition of new genomes and thus the analyses would be only slightly skewed. Analysis of the *core* genome showed that it retained the CDSs of the general physiological and survival functions of the species. In a previous study with only seven *B. thuringiensis* strains, the proportion of genes shared among the strains varied within a smaller range (59.0% to 68.2%) than that observed by us (49.3% to 71.1%) (FANG et al., 2011). Figure 5 shows the sharing of CDSs of BAC3151 with the other *B. thuringiensis*. Our isolate has more CDSs in common with MC28 (4951 genes) and less with Al Hakam (3965 CDSs). This observation is consistent with the greater proximity of BAC3151 to MC28 observed in the phylogenetic analysis. The lower number of genes shared with Al Hakam may be a result not only of a less close relationship, but also because Al Hakam has a smaller gene repertoire, thus decreasing the possibility of sharing.

In addition, there was a large variation in the number of *singletons* among *B. thuringiensis* strains, as reported by Fang et al. (2011). The *singletons* terminology describes the CDSs found in only one of the strains in our study. The same CDSs may perhaps be found in other strains or species. *B. thuringiensis* BAC3151 showed 221 *singletons*, about 4% of the total CDSs, whereas, in the other genomes, this value ranged from approximately 0.4% to 11% (for serovar *chinensis* CT-43 and for Bt407, respectively). Most *singletons* were associated with hypothetical proteins, as observed in other strains and species (SMITS et al., 2010; JOSEPH et al., 2012). The large number of hypothetical proteins found in various lineages may be the result of genes from horizontal transfer, including from phylogenetically distant organisms, for which molecular characterization has not yet been done. Indeed, *B. thuringiensis* can easily acquire genetic material from other bacteria, as reported previously (GONZALEZ et al., 1982; HAACK et al.,1996; VILAS-BÔAS et al.,1998). The acquisition of external genes may contribute to the better adaptation of bacterial subpopulations to specific niches (LAING et al., 2010). Other *singletons* found were mainly involved in gene regulation, synthesis of cellular components, stimulus response, and processing of genetic information (replication and repair).

(A)

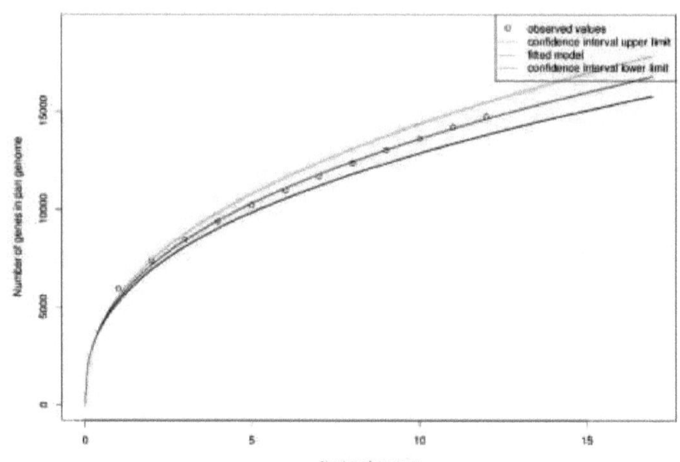
observed values
confidence interval upper limit
fitted model
confidence interval lower limit
Number of genes in pan genome
Number of genomes

(B)

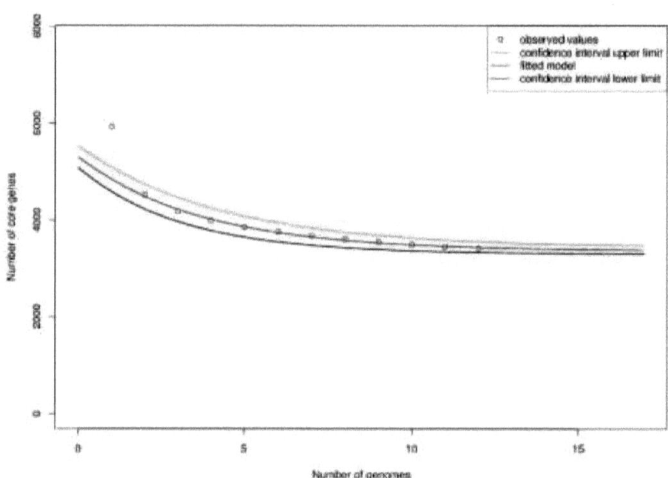
observed values
confidence interval upper limit
fitted model
confidence interval lower limit
Number of core genes
Number of genomes

**(C)**

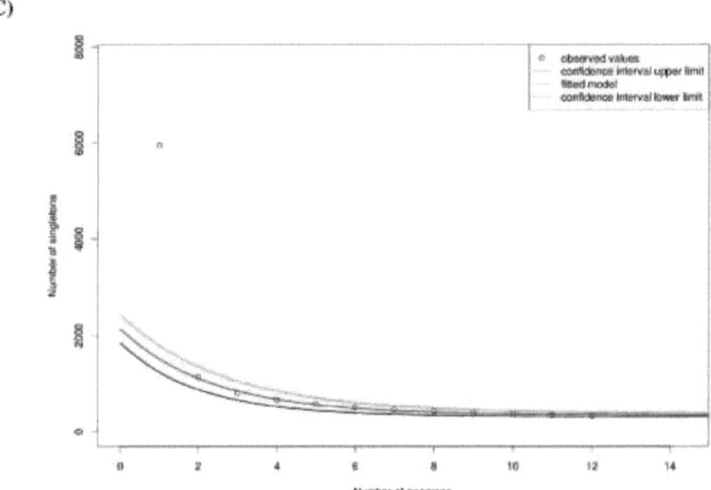

Figure 4. development of the pangenome, genome *core* and *singletons of B. thuringiensis* with increasing number of genomes. Nonlinear least squares fit models predict pangenoma development (A), genomic *core* (B) and singletons (C) of the genomes based on the functions $5422.261x^{0.299}$, $1935.165e^{(x^2.464)} + 3367.738$ and $1793.351e^{(x^2.465)} + 339.683$, respectively.

**(A)**           **(B)**

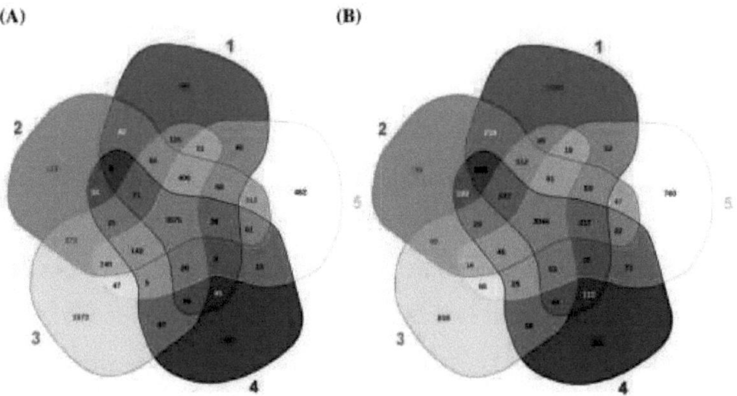

(C)

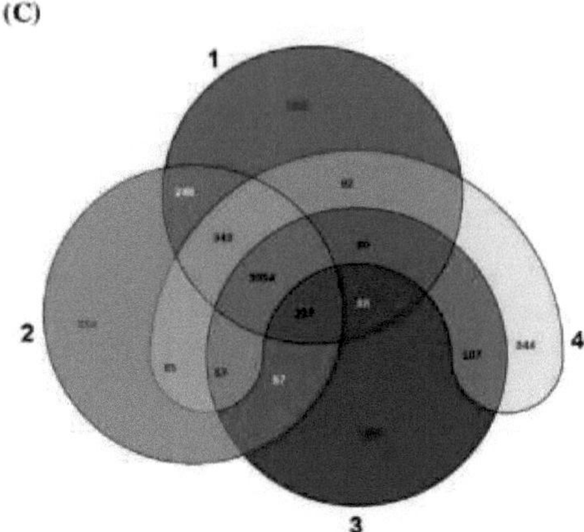

Figure 5. Venn diagrams constructed with EDGAR (BLOM et al., 2009) showing the sharing of CDSs of BAC3151 with other genomes. (A) 1. BMB171, 2. MC28, 3. sv. *thuringiensis* IS5056, 4. Al Hakam, 5. BAC3151; (B) 1. YBT-1518, 2. sv. *chinensis* CT-43, 3. YBT-020, 4. sv. *kurstaki* HD73, 5. BAC3151; (C) 1. Bt407, 2. HD-789, 3. sv. *konkukian* 97-27, 4. BAC3151. Overlapping regions represent common CDSs among the genomes analyzed.

## 4.5. Putative genes of *B. thuringiensis* BAC3151 associated with antagonism

Microorganisms compete for limited resources within the community and mechanisms of competition include antagonistic effects. *B. thuringiensis* BAC3151 has demonstrated *in vitro* antimicrobial activity against Gram-negative and Gram-positive bacteria, including *Curtobacterium flaccumfaciens* pv. *flaccumfaciens,* which is responsible for the bacterial wilt of bean (LOPES et al., 2015). Many secondary metabolite gene *clusters as* well as genes for other putative products, which may be related to antimicrobial activity, were identified in the *draft* genome of BAC3151. The major genes are shown in Table 5 and are potentially involved in the synthesis of bacteriocins, siderophores, polyketides, chitinases, N-acyl homoserine lactonase, insecticidal proteins and antibiotic compounds. These genes further point to possibilities for new tests that could expand the potential use of BAC3151 for biological control of plant diseases.

Table 5. Principal of key putative gene classes potentially involved in antagonist activity of *B. thuringiensis* BAC3151.

| Class | Size (bp) | Product | Coverage | *E-value* | Similarity |
|---|---|---|---|---|---|
| **Bacteriocins** | | | | | |
| scaffold3_orf00655 | 68,0 | Bacteriocin Biosynthesis Protein | 0,985294 | 2,00E-30 | 1,0 |
| scaffold21_orf00106 | 1435,0 | Peptidase from bacteriocin S8 | 0,987456 | 0,0 | 0,989422 |
| scaffold26_orf00190 | 649,0 | Bacteriocin biosynthesis protein (SagD) | 0,998459 | 0,0 | 0,978428 |
| scaffold29_orf00015 | 942,0 | Lantibiotic Biosynthesis Protein (LanM) | 0,998938 | 0,0 | 0,953291 |
| **Siderophores** | | | | | |
| scaffold19_orf00032 | 253,0 | Trilactone hydrolase (IroE) | 0,996047 | 1,00E-136 | 0,948617 |
| scaffold24_orf00456 | 612,0 | Rhizobactin Biosynthesis Protein (RhbC) | 0,998366 | 0,0 | 1,0 |
| scaffold24_orf00453 | 327,0 | Petrobactin Biosynthesis Protein (AsbE) | 0,996942 | 0,0 | 1,0 |
| scaffold24_orf00457 | 602,0 | Petrobactin Biosynthesis Protein (AsbA) | 0,998339 | 0,0 | 0,975083 |
| **Kinases** | | | | | |
| scaffold1_orf00008 | 235,0 | Chitinase C | 0,995745 | 1,00E-131 | 1,0 |

Table 5. Continued.

| scaffold19_orf00019 | 360,0 | Quitinase | 0,997222 | 0,0 | 0,977778 |
|---|---|---|---|---|---|
| scaffold28_orf00015 | 688,0 | Chitinase C | 0,986919 | 0,0 | 0,954412 |
| **Polyketides** | | | | | |
| scaffold9_orf00043 | 294,0 | Polyketide synthase | 0,996599 | 1,00E-171 | 1,0 |
| scaffold16_orf00424 | 47,0 | Polyketide synthase | 0,978723 | 1,00E-17 | 1,0 |

| Insecticide Proteins | | | | | |
| --- | --- | --- | --- | --- | --- |
| C1127_orf00006 | 1584,0 | Mosquitocidal Protein | 0,985 | 0,0 | 0,95 |
| scaffold24_orf00006 **Lactonase** | 292,0 | 5-endotoxin | 0,996575 | 1,00E-153 | 0,94863 |
| scaffold3_orf00391 | 250,0 | N-acyl homoserine lactonase | 0,996 | 1,00E-143 | 1,0 |
| **Antibiotics** | | | | | |
| scaffold24_orf00119 | 206,0 | Monooxygenase of antibiotic biosynthesis | 0,995146 | 1,00E-119 | 1,0 |
| scaffold26_orf00002 | 311,0 | Monooxygenase of antibiotic biosynthesis | 0,649518 | 1,00E-106 | 0,995074 |
| scaffold32_orf00043 | 492,0 | Monooxygenase of antibiotic biosynthesis | 0,457317 | 1,00E-108 | 0,884956 |
| scaffold32_orf00063 | 108,0 | Monooxygenase of antibiotic biosynthesis | 0,990741 | 3,00E-55 | 1,0 |

Bacteriocins have been well described for *Bacillus,* especially for *B. subtilis,* although strains of *B. thuringiensis* also have the potential to produce them (READ et al., 2009). Although several bacteriocins from the *B. cereus* group have been reported, relatively few have been lantibiotic. In BAC3151, a lantibiotic *cluster* was found to be highly similar to the turicin biosynthetic *cluster* (Figure 6A). Turicin production has been described in different *B. thuringiensis* strains, such as turicin B439 from *B. thuringiensis* B439 (AHERN et al., 2003), turicin 17 from *B. thuringiensis* NEB17 (LEE et ai., 2009b),

turicin H from *B. thuringiensis* SF361 (LEE et al., 2009a) and turicin S from *B. thuringiensis* sv. *entomocidus* HD198 (CHEHIMI et al., 2007).

This bacteriocin has demonstrated antibacterial activity mainly against Gram-positive bacteria (FAVRET and YOUSTEN, 1989; REA et al., 2014). *Bacillus* bacteriocins are increasingly important due to their inhibition spectrum that is sometimes broader than that of many lactic acid bacteria and can include Gram-negative bacteria, yeasts and other fungi, in addition to Gram-positive bacteria (ABRIOUEL et al., 2011).

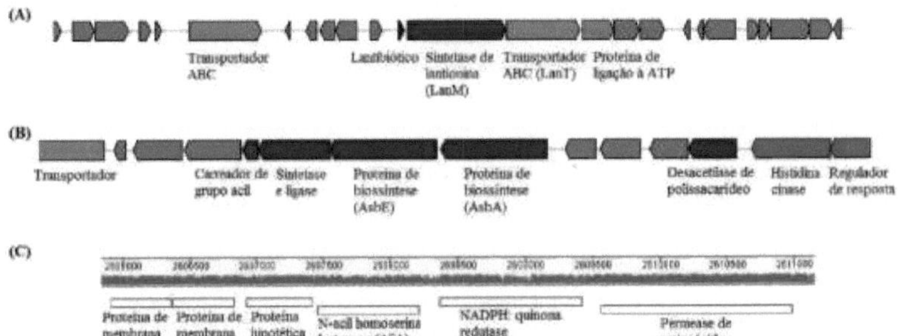

Figure 6. Some genomic regions of *B. thuringiensis* BAC3151 containing putative genes associated with microbial antagonism. (A) Lantibiotic gene *cluster* highly similar to turicin. (B) Petrobactin biosynthetic *cluster.* (C) Region containing N-acyl homoserine lactonase gene, presumably involved in *quorum sensing* inhibition by BAC3151. In (A) and (B), biosynthesis, transport and regulatory genes are represented in red, blue and green, respectively. Other genes are represented in gray. In (C), the profile in purple indicates the level of conservation of the region after alignment with other *B. thuringiensis* genomes.

Putative genes of BAC3151 involved in siderophore biosynthesis include the genes present in the petrobactin biosynthetic *cluster* (Figure 6B). Petrobactin production has been reported in *Bacillus* species, such as *B. thuringiensis* (WILSON et al., 2006), *B. cereus* (PARK et al., 2005), and *Bacillus anthracis* (CENDROWSKI et al., 2004). Siderophores of *B. thuringiensis,* such as petrobactin, may be especially important for biocontrol in agriculture, since *B. thuringiensis* is naturally associated with plants and competition for iron can lead to inhibition of phytopathogens, as well as increasing plant supply of the element. Phytopathogens are generally not able to produce comparable iron acquisition systems (RADDADI et al., 2007).

Another class of genes potentially involved in biocontrol is chitinases. Three putative chitinase genes (*chi* genes) have been identified in the genome. In *B. thuringiensis,* up to seven copies have been reported (MARCON et al., 2014). *Bacillus* spp. capable of degrading chitin in soil (KUZU et al., 2012), phylloplane (SMITT and COUCHE, 1991) and in association with insects (SNEH et al., 1993) have been described. Since chitin is an important component of the cell wall of fungi, chitinases can be used for the control of phytopathogenic fungi, protecting plants against diseases (TANG et al., 2012). Interestingly, the chitinolytic activity of *B. thuringiensis* may exhibit synergistic effect on insecticide toxicity. This can be applied to the development of better insecticidal products by combining both factors in a strain or in formulations (WIWAT et al., 2000; LIU et al., 2002).

In addition to these, polyketide synthase genes were found. Due to the versatile mechanism of synthesis of polyketides, they can exhibit diverse structures with a variety of biological functions, including antimicrobial activity. There is little information in the literature on polyketides from *B. thuringiensis*. However, the increasing number of genomes available for the species offers the opportunity for mining genes potentially involved in their synthesis and whose applicability could be explored. Polyketides, in fact, are difficult to obtain, which makes heterologous expression an alternative for use. Heterologous production systems may also facilitate the study of the catalytic properties of polyketide synthases. Examples of heterologous hosts for production of these proteins include *Escherichia coli, B. subtilis,* and *Streptomyces coelicolor* (RUDE and KHOSLA, 2004).

Two insecticidal genes were also found in BAC3151. One (C1127_orf00006) with 95% similarity to a mosquitocidal protein gene from *B. thuringiensis* IBL 4222, and another (scaffold24_orf00006) with 94.7% similarity to an insecticidal ô-endotoxin from *B. thuringiensis* MC28. The number of insecticidal genes varies among strains (FANG et al., 2011), which can produce one or more parasporal protein crystals. The individual insecticidal proteins that constitute the crystals usually have a spectrum of toxic activity restricted to a few species of a particular insect order, mainly lepidopterans, dipterans, coleopterans and hymenopterans (RHO et al., 2007). Activity against nematodes and other invertebrates has also been described (WEI et al., 2003; KOTZE et al., 2005; RHO et al., 2007). The combination of toxins from a given strain of *B. thuringiensis* defines the spectrum of activity of that strain. However, there are many crystal-producing strains with no known toxic activity (RHO et al., 2007). Further analysis may confirm the production of parasporal crystals by BAC3151 and its toxicity, as advances in the development of new biopesticides based on *B. thuringiensis* and the manipulation of its genes depend on the variety of available strains and their corresponding insecticidal genes.

Regarding the gene encoding N-acyl homoserine lactonase (*aiiA* gene*)* (Figure 6C), it is presumably responsible for the observed inhibition of *quorum sensing of* Gram-negative bacteria by BAC3151 (LOPES et al., 2015). Many Gram-negative phytopathogenic bacteria use quorum *sensing* as a mechanism for regulating different biological activities, including the production of virulence factors (VON BODMAN et al., 2003). The *aiiA* has been found in different strains of *B. thuringiensis* (LEE et al., 2002) and inhibition of chemical communication of phytopathogenic bacteria may be an efficient strategy for plant disease control. Studies indicate that the use of *quorum sensing* inhibitors for this purpose is promising, extending the current ways of preventing bacterial diseases (DONG et al., 2004).

Other genes, especially those of monooxygenases, that possibly participate in the synthesis of antibiotic compounds were found in the genome of *B. thuringiensis* BAC3151. These genes are often part of *clusters* involved in a complex secondary metabolism characteristic of strains producing diverse antimicrobials. Although the fraction of the BAC3151 genome assigned to antimicrobial synthesis is apparently not as large as that of some *Bacillus* spp. such as B. *subtilis* and *B. amyloliquefaciens* (KUNST et al., 1997; CHEN et al., 2009), future research should analyze the functionality of the genes found, the conservation of the synthesis pathways, and the properties of the eventual products as well as their applicability in agriculture or even in other areas. Thus, the isolate may represent a source of secondary metabolites and important enzymes to control pathogenic microorganisms.

# CHAPTER 5

## 5. CONCLUSIONS

In this work, it was found that *B. thuringiensis* BAC3151 endophytic from common bean was more similar to *B. thuringiensis* MC28 soil isolate. Genomic analyses also revealed several genes from BAC3151 without significant similarity to others in the public databases as well as genes not found in the other strains analyzed and that will be further investigated in the future. In addition, putative genes involved in multiple forms of antagonism are present. In summary, the results presented from the genomic analyses may contribute to the development of new strategies for the biological control of diseases of bean and other crops.

## 6. REFERENCES

ABDERRAHMANI, A.; TAPI, A.; NATECHE, F.; CHOLLET, M.; LECLERE, V.; WATHELET, B.; et al. Bioinformatics and molecular approaches to detect NRPS genes involved in the biosynthesis of kurstakin from *Bacillus thuringiensis*. **Appl Microbiol Biotechnol**, v. 92, p. 571-581, 2011.

ABRIOUEL, H.; FRANZ, C. M.; BEN OMAR, N.; GALVEZ, A. Diversity and applications of *Bacillus* bacteriocins. **FEMS Microbiol Rev**, v. 35, p. 201-232, 2011.

AHERN, M.; VERSCHUEREN, S.; VAN SINDEREN, D. Isolation and characterisation of a novel bacteriocin produced by *Bacillus thuringiensis* strain B439. **FEMS Microbiol Lett**, v. 220, p. 127-131, 2003.

AKIBA, Y. Microbial ecology of *Bacillus thuringiensis*. **Appl Ent Zool**, v. 21, p. 76-80, 1986.

ALKAN, C.; SAJJADIAN, S.; EICHLER, E. E. Limitations of next-generation genome sequence assembly. **Nat Methods**, v. 8, p. 61-65, 2011.

ALTSCHUL, S. F.; GISH, W.; MILLER, W.; MYERS, E. W.; LIPMAN, D. J. Basic local alignment search tool. **J Mol Biol**, v. 215, p. 403-410, 1990.

ANDREOLLI, M.; LAMPIS, S.; POLI, M.; GULLNER, G.; BIRO, B.; VALLINI, G. Endophytic *Burkholderia fungorum* DBT1 can improve phytoremediation efficiency of polycyclic aromatic hydrocarbons. **Chemosphere**, v. 92, p. 688-694, 2013.

ANSARI, M. Z.; YADAV, G.; GOKHALE, R. S.; MOHANTY, D. NRPS- PKS: a knowledge-based resource for analysis of NRPS/PKS megasynthases. **Nucleic Acids Res**, v. 32, p. W405-413, 2004.

ARANTES, O. M. N.; VILAS-BÔAS, L. A.; VILAS-BÔAS, G. F. L. T. *Bacillus thuringiensis:* strategies in biological control. In: SERAFINI, L. A.; BARROS, N. M.; AZEVEDO, J. L. (org). **Biotecnologia na agricultura e agroindústria**, Guaíba: Agropecuária, v. 2, p. 269-293, 2002.

ARGUELLES-ARIAS, A.; ONGENA, M.; HALIMI, B.; LARA, Y.; BRANS, A.; JORIS, B.; et al. *Bacillus amyloliquefaciens* GA1 as a source of potent antibiotics and other secondary metabolites for biocontrol of plant pathogens. **Microb Cell Fact**, v. 8, p. 63, 2009.

ARONSON, A. I.; BECKMAN, W.; DUNN, P. *Bacillus thuringiensis* and related insect pathogens. **Microbiol Rev**, v. 50, p. 1-24, 1986.

BAKER, M. *De novo* genome assembly: what every biologist should know. **Nat Meth**, v. 9, p. 333-337, 2012.

BAKKER, P. A. H. M.; PIETERSE, C. M. J.; VAN LOON, L. C. Induced systemic resistance by fluorescent *Pseudomonas* spp. **Phytopathology**, v. 97, p. 239243, 2007.

BANSODE, V. B.; BAJEKAL, S. S. Characterization of chitinases from microorganisms isolated from Lonar Lake. **Ind J Biotech**, v. 5, p. 357-363, 2006.

BARBOZA-CORONA, J. E.; VAZQUEZ-ACOSTA, H.; BIDESHI, D. K.; SALCEDO-HERNANDEZ, R. Bacteriocin-like inhibitor substances produced by Mexican strains of *Bacillus thuringiensis*. **Arch Microbiol**, v. 187, p. 117-126, 2007.

BARKA, E. A.; GOGNIES, S.; NOWAK, J.; AUDRAN, J. C.; BELARBI, A. Inhibitory effect of endophyte bacteria on *Botrytis cinerea* and its influence to promote the grapevine growth. **Biol Control**, v. 24, p. 135-142, 2002.

BAVYKIN, S. G.; LYSOV, Y. P.; ZAKHARIEV, V.; KELLY, J. J.; JACKMAN, J.; STAHL, D. A.; et al. Use of 16S rRNA, 23S rRNA, and *gyrB* gene sequence analysis to determine phylogenetic relationships of *Bacillus cereus* group microorganisms. **J Clin Microbiol**, v. 42, p. 3711-3730, 2004.

BENDTSEN, J. D.; NIELSEN, H.; VON HEIJNE, G.; BRUNAK, S. Improved prediction of signal peptides: SignalP 3.0. **J Mol Biol**, v. 340, p. 783-795, 2004.

BERLINER, E. Eber die schlaffsucht der Mehlmottenraupe. *(Ephesis kuehniella* Zell.) undihren Erreger *Bacillus thuringiensis,* n. sp. Z. ang. **Entomology**, v. 2, p. 29-56, 1915.

BERLINER, E. Über die schlafsucht der Mehlmottenraupe. **Z. Gesante Getreidewesen**, v. 3, p. 63-70, 1911.

BHATTACHRYA, D.; NAGPURE, A.; GUPTA, R. K. Bacterial

chitinase: properties and potential. **Critic Rev Biotechnol**, v. 27, p. 21-28, 2007.

BLOM, J.; ALBAUM, S. P.; DOPPMEIER, D.; PUHLER, A.; VORHOLTER, F. J.; ZAKRZEWSKI, M; et al. EDGAR: a software framework for the comparative analysis of prokaryotic genomes. **BMC Bioinformatics**, v. 10, p. 154, 2009.

BODE, H.B. Entomopathogenic bacteria as a source of secondary metabolites. **Curr Opin Chem Biol**, v. 13, p. 2224-230, 2009.

BRADER, G.; COMPANT, S.; MITTER, B.; TROGNITZ, F.; SESSITSCH, A. Metabolic potential of endophytic bacteria. **Curr Opin Biotechnol**, v. 27, p. 3037, 2014.

BRAR, S. K.; VERMA, M.; TYAGI, R. D.; VALÉRO, J. R. Recent advances in downstream processing and formulations of *Bacillus thuringiensis* based biopesticides. **Process Biochem**, v. 41, p. 323-342, 2006.

BRAVO, A.; SARABIA, S.; LOPEZ, L.; ONTIVEROS, H.; ABARCA, C.; ORTIZ, A.; et al. Characterization of *cry* genes in a Mexican *Bacillus thuringiensis* strain collection. **Appl Environ Microbiol**, v. 64, p. 4965-4972, 1998.

CÁRDENAS, M. I.; GALÁN-WONG, L.; FERRÉ-MANZANERO, J.; PEREYRA-ALFÉREZ, B. Selección de toxinas *cry* contra *Thichoplusia ni* . **Cienc Uanl**, v. 4, p. 51-62, 2001.

CARLSON, C. R.; CAUGANT, D. A.; KOLSTO, A. B. Genotypic diversity among *Bacillus cereus* and *Bacillus thuringiensis* strains. **Appl Environ Microbiol**, v. 60, p. 1719-1725, 1994.

CARLSON, C. R.; JOHANSEN, T.; KOLSTO, A. B. The chromosome map of *Bacillus thuringiensis* subsp. *canadensis* HD224 is highly similar to that of the *Bacillus cereus* type strain ATCC 14579. **FEMS Microbiol Lett**, v. 141, p. 163-167, 1996.

CASTRESANA J. Selection of conserved blocks from multiple alignments for their use in phylogenetic analysis. **Mol Biol Evol**, v. 17, p. 540-552, 2000.

CENDROWSKI, S.; MACARTHUR, W.; HANNA, P. *Bacillus anthracis* requires siderophore biosynthesis for growth in macrophages and mouse virulence. **Mol Microbiol**, v. 51, p. 407-417, 2004.

CHAISSON, M.; PEVZNER, P. Short read fragment assembly of bacterial genomes. **Genome Res**, v. 18, p. 324, 2008.

CHALLACOMBE, J. F.; ALTHERR, M. R.; XIE, G.; BHOTIKA, S. S.; BROWN, N.; BRUCE, D.; et al. The complete genome sequence of *Bacillus thuringiensis* Al Hakam. **J Bacteriol**, v. 189, p. 3680-3681, 2007.

CHEHIMIMI, S.; DELALANDE, F.; SABLE, S.; HAJLAOUI, M. R.; VAN DORSSELAER, A.; LIMAM, F.; et al. Purification and partial amino acid sequence of thuricin S, a new *anti-Listeria* bacteriocin from *Bacillus thuringiensis*. **Can J Microbiol**, v. 53, p. 284-290, 2007.

CHEN, X. H.; KOUMOUTSI, A.; SCHOLZ, R.; EISENREICH, A.; SCHNEIDER, K.; HEINEMEYER, I.; et al. Comparative analysis of the complete genome sequence of the plant growth-promoting bacterium *Bacillus amyloliquefaciens* FZB42. **Nat Biotechnol**, v. 25, p. 1007-1014, 2007.

CHEN, X. H.; KOUMOUTSI, A.; SCHOLZ, R.; SCHNEIDER, K.; VATER, J.; SUSSMUTH, R.; et al. Genome analysis of *Bacillus amyloliquefaciens* FZB42 reveals its potential for biocontrol of plant pathogens. **J Biotechnol**, v. 140, p. 27-37, 2009.

CHERIF, A.; CHEHIMI, S.; LIMEM, F.; HANSEN, B. M.; HENDRIKSEN, N. B.; DAFFONCHIO, D.; et al. Detection and characterization of the novel bacteriocin *entomocidus* 9, and safety evaluation of its producer, *Bacillus thuringiensis* ssp. *entomocidus* HD9. **J Appl Microbiol**, v. 95, p. 990-1000, 2003.

CHERIF, A.; REZGUI, W.; RADDADI, N.; DAFFONCHIO, D.; BOUDABOUS, A. Characterization and partial purification of entomocin 110, a newly identified bacteriocin from *Bacillus thuringiensis* subsp. *entomocidus* HD110. **Microbiol Res**, v. 163, p. 684-692, 2008.

CHEVREUX, B.; PFISTERER, T.; DRESCHER, B.; DRIESEL, A. J.; MULLER, W. E.; WETTER, T.; et al. Using the miraEST assembler for reliable and automated mRNA transcript assembly and SNP detection in sequenced ESTs. **Genome Res**, v. 14, p. 1147-1159, 2004.

COOMBS, J. T.; MICHELSEN, P. P.; FRANCO, C. M. M. Evaluation of endophytic actinobacteria as antagonists of *Gaeumannomyces graminis* var. *tritici* in wheat. **Biol Control**, v. 29, p. 359-366, 2004.

COSTA, L. E. O.; QUEIROZ, M. V.; BORGES, A. C.; MORAES, C. A.; ARAÚJO, E. F. Isolation and characterization of endophytic bacteria isolated from the leaves of the common bean *(Phaseolus vulgaris)*. **Braz J Microbiol**, v. 43, p. 1562-1575, 2012.

CRICKMORE, N.; BAUM, J.; BRAVO, A.; LERECLUS, D.; NARVA, K.; SAMPSON, K.; et al. ***Bacillus thuringiensis* toxin nomenclature**. Available at: http://www.btnomenclature.info/. Accessed on: December 20, 2014.

DAHIYA, N.; TEWARI, R.; HOONDAL, G. S. Biotechnological aspects of chitinolytic enzymes: a review. **Appl Microbiol Biotechnol**, v. 71, p. 773-782, 2006.

D'ALESSANDRO, M.; ERB, M.; TON, J.; BRANDENBURG, A.; KARLEN, D.; ZOPFI, J.; et al. Volatiles produced by soil-borne endophytic bacteria increase plant pathogen resistance and affect tritrophic interactions. **Plant Cell Environ**, v. 37, p. 813-826, 2014.

DARLING, A. C.; MAU, B.; BLATTNER, F. R.; PERNA, N. T. Mauve: multiple alignment of conserved genomic sequence with rearrangements. **Genome Res**, v. 14, p. 1394-1403, 2004.

DE LA FUENTE-SALCIDO, N.; GUADALUPE ALANIS-GUZMAN, M.; BIDESHI, D. K.; SALCEDO-HERNANDEZ, R.; BAUTISTA-JUSTO, M.; BARBOZA-CORONA, J. E. Enhanced synthesis and antimicrobial activities of bacteriocins produced by Mexican strains of *Bacillus thuringiensis*. **Arch Microbiol**, v. 190, p. 633-640, 2008.

DE LA VEGA, L. M.; BARBOZA-CORONA, J. E.; AGUILAR-USCANGA, M. G.; RAMIREZ-LEPE, M. Purification and characterization of an exochitinase from *Bacillus thuringiensis* subsp. *aizawai* and its action against phytopathogenic fungi. **Can J Microbiol**, v. 52, p. 651-657, 2006.

DE MAAGD, R. A.; BOSCH, D.; STIEKEMA, W. *Bacillus thuringiensis* toxin-mediated insect resistance in plants. **Trends Plant Sci**, v. 4, p. 9-13, 1999.

DELCHER, A. L.; HARMON, D.; KASIF, S.; WHITE, O.; SALZBERG, S. L. Improved microbial gene identification with GLIMMER. **Nucleic Acids Res**, v. 27, p. 4636-4641, 1999.

DIDELOT, X.; FALUSH, D. Inference of bacterial microevolution using multilocus sequence data. **Genetics**, v. 175, p. 1251-1266, 2007.

DOGGETT, N. A.; STUBBEN, C. J.; CHERTKOV, O.; BRUCE, D. C.; DETTER, J. C.; JOHNSON, S. L.; HAN, C. S. Complete genome sequence of *Bacillus thuringiensis* serovar *israelensis* strain HD-789. **Genome Announc**, v. 1, p. 1-2, 2013.

DOHM, J. C.; LOTTAZ, C.; BORODINA, T.; HIMMELBAUER, H. SHARCGS, a fast and highly accurate short-read assembly algorithm for de novo genomic sequencing. **Genome Res**, v. 17, p. 1697-1706, 2007.

DONG, Y. H.; ZHANG, X. F.; XU, J. L.; ZHANG, L. H. Insecticidal *Bacillus thuringiensis* silences *Erwinia carotovora* virulence by a new form of microbial antagonism, signal interference. **Appl Environ Microbiol**, v. 70, p. 954-960, 2004.

DOTY, S. L. Enhancing phytoremediation through the use of transgenics and endophytes. **New Phytol,** v. 179, p. 318-333, 2008.

DWORZANSKI, J. P.; DICKINSON, D. N.; DESHPANDE, S. V.; SNYDER, A. P.; ECKENRODE, B. A. Discrimination and phylogenomic classification of *Bacillus anthracis-cereus-thuringiensis* strains based on LC-MS/MS analysis of whole cell protein digests. **Anal Chem**, v. 82, p. 145-155, 2010.

EDGAR, R. C. MUSCLE: multiple sequence alignment with high accuracy and high throughput. **Nucleic Acids Res**, v. 32, p. 1792-1797, 2004.

EICHLER, E. E.; SANKOFF, D. Structural dynamics of eukaryotic chromosome evolution. **Science**, v. 301, p. 793-797, 2003.

EKBLOM, R.; WOLF, J. B. A field guide to whole-genome sequencing, assembly and annotation. **Evol Appl**, v. 7, p. 1026-1042, 2014.

FANG, Y.; LI, Z.; LIU, J.; SHU, C.; WANG, X.; ZHANG, X.; et al. A pangenomic study of *Bacillus thuringiensis*. **J Genet Genomics**, v. 38, p. 567-576, 2011.

FAVRET, M. E.; YOUSTEN, A. A. Thuricin: The bacteriocin produced by *Bacillus thuringiensis*. **J Invertebr Pathol**, v. 53, p. 206-216, 1989.

FEDERICI, B. A. *Bacillus thuringiensis* in Biological Control. In: BELLOWS, T.; FISHER, T. W. (eds). **Handbook of Biological Control.**

California: University of California, p. 575-593, 1999.

FEITELSON, J. S. Novel pesticidal delta-endotoxins from *Bacillus thuringiensis*. In: **Proceedings of XXVII Annual Meeting of the Society for Invertebrate Pathology**, Montpellier, France, p. 184, 1994.

FELNAGLE, E. A.; JACKSON, E. E.; CHAN, Y. A.; PODEVELS, A. M.; BERTI, A. D.; MCMAHON, M. D.; et al. Nonribosomal peptide synthetases involved in the production of medically relevant natural products. **Mol Pharm**, v. 5, p. 191-211, 2008.

FELSENSTEIN, J. PHYLIP: Phylogeny Inference Package (version3.2). **Cladistics**, v. 5, p. 164-166, 1989.

FERREIRA, L. H. P. L.; SUZUKI, M. T.; ITANO, E. N.; ONO, M. A.; ARANTES, O. M. N. Ecological aspects of *Bacillus thuringiensis* in an Oxisol. **Sci Agric**, v. 60, p.19-22, 2003.

FICKERS, P. Antibiotic compounds from *Bacillus:* Why are they so amazing? **Am J Biochem Biotechnol**, v. 8, p. 38-43, 2012.

FLEISCHMANN, R. D.; ADAMS, M. D.; WHITE, O.; CLAYTON, R. A.; KIRKNESS, E. F.; KERLAVAGE, A. R.; et al. Whole-genome random sequencing and assembly of *Haemophilus influenzae* Rd. **Science**, v. 269, p. 496-512, 1995.

FRASER, C. M.; GOCAYNE, J. D.; WHITE, O.; ADAMS, M. D.; CLAYTON, R. A.; FLEISCHMANN, R. D.; et al. The minimal gene complement of *Mycoplasma genitalium*. **Science**, v. 270, p. 397-403, 1995.

FROMMEL, M. I.; NOWAK, J.; LAZAROVITS, G. Growth enhancement and developmental modifications of *in vitro* grown potato (*Solanum tuberosum* spp. *tuberosum)* as affected by a nonfluorescent *Pseudomonas* sp. **Plant Physiol**, v. 96, p. 928-936, 1991.

FROST, L. S.; LEPLAE, R.; SUMMERS, A. O.; TOUSSAINT, A. Mobile genetic elements: the agents of open source evolution. **Nat Rev Microbiol**, v. 3, p. 722-732, 2005.

FUQUA, C.; PARSEK, M. R.; GREENBERG, E. P. Regulation of gene expression by cell-to-cell communication: acyl-homoserine lactone quorum sensing. **Annu Rev Genet**, v. 35, p. 439-468, 2001.

GALLEGOS, M. T.; SCHLEIF, R.; BAIROCH, A.; HOFMANN, K.; RAMOS, J. L. Arac/XylS family of transcriptional regulators. **Microbiol**

**Mol Biol Rev**, v. 61, p. 393-410, 1997.

GARDNER, J. M.; FELDMAN, A. W.; ZABLOTOWICZ, R. M. Identity and behavior of xylem-residing bacteria in rough lemon roots of florida citrus trees. **Appl Environ Microbiol**, v. 43, p. 1335-1342, 1982.

GELERNTER, W.; SCHWAB, G. E. Transgenic bacteria, viruses, algae and other microorganisms as *Bacillus thuringiensis* toxin delivery systems. In: ENTWISTLE, P. F.; CORY, J. S.; BAILEY, M. J.; HIGGS, S. (ed). ***Bacillus thuringiensis*, an environmental biopesticide: theory and practice**, Chichester: John Wiley & Sons, p. 89-104, 1993.

GISH, W.; STATES, D. J. Identification of protein coding regions by database similarity search. **Nat Genet**, v. 3, p. 266-272, 1993.

GLARE, T. R; O'CALLAGHAN, M. *Bacillus thuringiensis:* biology, ecology and safety. In: ENTWISTLE, P. F.; CORY, J. S.; BAILEY, M. J.; HIGGS, S. (ed). ***Bacillus thuringiensis*, an environmental biopesticide: theory and practice**, Chichester: John Wiley, p. 350, 2000.

GOMAA, E. Z. Chitinase production by *Bacillus thuringiensis* and *Bacillus licheniformis:* their potential in antifungal biocontrol. **J Microbiol**, v. 50, p. 103-111, 2012.

GONZALEZ, J. M., JR.; BROWN, B. J.; CARLTON, B. C. Transfer of *Bacillus thuringiensis* plasmids encoding for delta-endotoxin among strains of *B. thuringiensis* and *B. cereus*. **Proc Natl Acad Sci USA**, v. 79, p. 6951-6955, 1982.

GORDON, D.; ABAJIAN, C.; GREEN, P. Consed: a graphical tool for sequence finishing. **Genome Res**, v. 8, p. 195-202, 1998.

GRIFFITHS-JONES, S.; MOXON, S.; MARSHALL, M.; KHANNA, A.; EDDY, S. R.; BATEMAN, A. Rfam: annotating non-coding RNAs in complete genomes. **Nucleic Acids Res**, v. 33, p. D121-124, 2005.

GUAN, P.; AI, P.; DAI, X.; ZHANG, J.; XU, L.; ZHU, J.; et al. Complete genome sequence of *Bacillus thuringiensis* serovar *sichuansis* strain MC28. **J Bacteriol**, v. 194, p. 6975, 2012.

GUNDOGDU, O.; BENTLEY, S. D.; HOLDEN, M. T.; PARKHILL, J.; DORRELL, N.; WREN, B. W. Re-annotation and re-analysis of the *Campylobacter jejuni* NCTC11168 genome sequence. **BMC Genomics**, v. 8, p. 162, 2007.

GUSTAFSON, K. R.; CARDELLINA, J. H., 2ND; FULLER, R. W.; WEISLOW, O. S.; KISER, R. F.; SNADER, K. M.; et al. AIDS-antiviral sulfolipids from cyanobacteria (blue-green algae). **J Natl Cancer Inst**, v. 81, p. 1254-1258, 1989.

HAACK, B. J.; ANDREWS, R. E.; LOYNACHAN, T. E. Tn916-mediated genetic exchange in soil. **Soil Biol Biochem**, v. 28, p. 765-771, 1996.

HALLMANN, J.; QUADT-HALLMANN, A.; MAHAFEE, W. F.; KLOEPPER, J. Bacterial endophytes in agricultural crops. **Can J Microbiol**, v. 43, p. 895-914, 1997.

HALLMANN, J.; QUADT-HALLMANN, A.; RODRIGUEZ-KABANA, R.; KLOEPPER, J. W., Interactions between *Meloidogyne incognita* and endophytic bacteria in cotton and cucumber. **Soil Biol Biochem**, v. 30, p. 925-937, 1998.

HAN, C. S.; XIE, G.; CHALLACOMBE, J. F.; ALTHERR, M. R.; BHOTIKA, S. S.; BROWN, N.; et al. Pathogenomic sequence analysis of *Bacillus cereus* and *Bacillus thuringiensis* isolates closely related to *Bacillus anthracis*. **J Bacteriol**, v. 188, p. 3382-3390, 2006.

HANNAY, C. L.; FITZ-JAMES, P. C. The protein-crystals of *Bacillus thuringiensis berliner*. **Can J Microbiol**, v. 1, p. 694-710, 1995.

HANSON, A. D.; PRIBAT, A.; WALLER, J. C.; DE CRECY-LAGARD, V. 'Unknown' proteins and 'orphan' enzymes: the missing half of the engineering parts list-and how to find it. **Biochem J**, v. 425, p. 1-11, 2010.

HARDOIM, P. R.; VAN OVERBEEK, L. S.; ELSAS, J. D. Properties of bacterial endophytes and their proposed role in plant growth. **Trends Microbiol**, v. 16, 463-471, 2008.

HAYES, M.; CARNEY, B.; SLATER, J.; BRUCK, W. Mining marine shellfish wastes for bioactive molecules: chitin and chitosan - Part A: extraction methods. **Biotechnol J**, v. 3, p. 871-877, 2008.

HE, J.; SHAO, X.; ZHENG, H.; LI, M.; WANG, J.; ZHANG, Q.; et al. Complete genome sequence of *Bacillus thuringiensis* mutant strain BMB171. **J Bacteriol**, v. 192, p. 4074-4075, 2010.

HE, J.; WANG, J.; YIN, W.; SHAO, X.; ZHENG, H.; LI, M.; et al.

Complete genome sequence of *Bacillus thuringiensis* subsp. *chinensis* strain CT-43. **J Bacteriol**, v. 193, p. 3407-3408, 2011.

HELGASON, E.; OKSTAD, O. A.; CAUGANT, D. A.; JOHANSEN, H. A.; FOUET, A.; MOCK, M.; et al. *Bacillus anthracis, Bacillus cereus, and Bacillus thuringiensis* - one species on the basis of genetic evidence. **Appl Environ Microbiol**, v. 66, p. 2627-2630, 2000.

HELGASON, E.; TOURASSE, N. J.; MEISAL, R.; CAUGANT, D. A.; KOLSTO, A. B. Multilocus sequence typing scheme for bacteria of the *Bacillus cereus* group. **Appl Environ Microbiol**, v. 70, p. 191-201, 2004.

HENDRIKSEN, N. B.; HANSEN, B. M. Long-term survival and germination of *Bacillus thuringiensis* var. *kurstaki* in a field trial. **Can J Microbiol**, v. 48, p. 256261, 2002.

HERTWECK, C.; LUZHETSKYY, A.; REBETS, Y.; BECHTHOLD, A. Type II polyketide synthases: gaining a deeper insight into enzymatic teamwork. **Nat Prod Rep**, v. 24, p. 162-190, 2007.

HUANG, J. Ultrastructure of bacterial penetration in plants. **Annu Rev Phytopathol**, v. 24, p. 141-157, 1986.

HUNT, M.; NEWBOLD, C.; BERRIMAN, M.; OTTO, T. D. A comprehensive evaluation of assembly scaffolding tools. **Genome Biol**, v. 15, p. R42, 2014.

HUNTER, S.; APWEILER, R.; ATTWOOD, T. K.; BAIROCH, A.; BATEMAN, A.; BINNS, D.; et al. InterPro: the integrative protein signature database. **Nucleic Acids Res**, v. 37, p. D211-215, 2009.

HUREK, T.; REINHOLD-HUREK, B.; VAN MONTAGU, M.; KELLENBERGER, E. Root colonization and systemic spreading of *Azoarcus* sp. strain BH72 in grasses. **J Bacteriol**, v. 176, 1913-1923, 1994.

HUSZ, B. On the use of *Bacillus thuringiensis* in the fight against the corner borer. **Corn Borer Instit Invest Sci Repub**, v.2, p. 99-110, 1929.

IBRAHIM, M. A.; GRIKO, N.; JUNKER, M.; BULLA, L. A. *Bacillus thuringiensis:* a genomics and proteomics perspective. **Bioeng Bugs**, v. 1, p. 31-50, 2010.

ISHIWATA, S. On a kind of severe flacherie (sotto disease). **Danihon Sanshi Kaiho**, v. 114, p. 1-5, 1901.

JENKE-KODAMA, H.; DITTMANN, E. Evolution of metabolic diversity: insights from microbial polyketide synthases. **Phytochemistry**, v. 70, p. 1858-1866, 2009.

JEONG, H.; PARK, S. H.; CHOI, S. K. Genome sequence of the acrystalliferous *Bacillus thuringiensis* serovar *israelensis* strain 4Q7, widely used as a recombination host. **Genome Announc**, v. 2, p. e00231-14, 2014.

JHA, P. N.; KUMAR, A. Endophytic colonization of *Typha australis* by a plant growth-promoting bacterium *Klebsiella oxytoca* strain GR-3. **J Appl Microbiol**, v. 103, p. 1311-1320, 2007.

JONES, P.; BINNS, D.; CHANG, H. Y.; FRASER, M.; LI, W.; MCANULLA, C.; et al. InterProScan 5: genome-scale protein function classification. **Bioinformatics**, v. 30, p. 1236-1240, 2014.

JOSEPH, S.; DESAI, P.; JI, Y.; CUMMINGS, C. A.; SHIH, R.; et al. Comparative analysis of genome sequences covering the seven *Cronobacter* species. **Plos One**, v. 7, p. e49455, 2012.

KIM, W.; HONG, Y. P.; YOO, J. H.; LEE, W. B.; CHOI, C. S.; CHUNG, S. I. Genetic relationships of *Bacillus anthracis* and closely related species based on variable-number tandem repeat analysis and BOX-PCR genomic *fingerprinting*. **FEMS Microbiol Lett**, v. 207, p. 21-27, 2002.

KLOEPPER, J. W.; RYU, C. M., ZHANG, S. Induced systemic resistance and promotion of plant growth by *Bacillus* spp. **Phytopathology**, v. 94, p. 1259-1266, 2004.

KOBAYASHI, T.; SUZUKI, M.; INOUE, H.; ITAI, R. N.; TAKAHASHI, M.; NAKANISHI, H.; et al. Expression of iron-acquisition-related genes in iron-deficient rice is co-ordinately induced by partially conserved iron-deficiency-responsive elements. **J Exp Bot**, v. 56, p. 1305-1316, 2005.

KOONIN, E. V. Computational genomics. **Curr Biol**, v. 11, p. R155-158, 2001.

KOTZE, A. C.; O'GRADY, J.; GOUGH, J. M.; PEARSON, R.; BAGNALL, N. H.; KEMP, D. H.; et al. Toxicity of Bacillus thuringiensis to parasitic and free-living life-stages of nematode parasites of livestock. **Int J Parasitol**, v. 35, p. 10131022, 2005.

KUNST, F.; OGASAWARA, N.; MOSZER, I.; ALBERTINI, A. M.; ALLONI, G.; AZEVEDO, V.; et al. The complete genome sequence of the grampositive bacterium *Bacillus subtilis*. **Nature**, v. 390, p. 249-256, 1997.

KUZU, S. B.; GUVENMEZ, H. K.; DENIZCI, A. A. Production of a thermostable and alkaline chitinase by *Bacillus thuringiensis* subsp. *kurstaki* strain HBK-51. **Biotechnol Res Int**, v. 2012,.p. 135498, 2012.

KYRPIDES, N. C. Fifteen years of microbial genomics: meeting the challenges and fulfilling the dream. **Nat Biotechnol**, v. 27, p. 627-632, 2009.

LAGESEN, K.; HALLIN, P.; RODLAND, E. A.; STAERFELDT, H. H.; ROGNES, T.; USSERY, D. W. RNAmmer: consistent and rapid annotation of ribosomal RNA genes. **Nucleic Acids Res**, v. 35, p. 3100-3108, 2007.

LAING, C.; BUCHANAN, C.; TABOADA, E. N.; ZHANG, Y.; KROPINSKI, A.; VILLEGAS, A.; et al. Pan-genome sequence analysis using Panseq: an online tool for the rapid analysis of core and accessory genomic regions. **BMC Bioinformatics**, v. 11, p. 461, 2010.

LAMB, T. G.; TONKYN, D. W.; KLUEPFEL, D. A. Movement of *Pseudomonas aureofaciens* from the rhizosphere to aerial plant tissue. **Can J Microbiol**, v. 42, p. 1112-1120, 1996.

LASLETT, D.; CANBACK, B. ARAGORN, a program to detect tRNA genes and tmRNA genes in nucleotide sequences. **Nucleic Acids Res**, v. 32, p. 11-16, 2004.

LEE, H.; CHUREY, J. J.; WOROBO, R. W. Biosynthesis and transcriptional analysis of thurincin H, a tandem repeated bacteriocin genetic locus, produced by *Bacillus thuringiensis* SF361. **FEMS Microbiol Lett**, v. 299, p. 205-213, 2009a.

LEE, K. D.; GRAY, E. J.; MABOOD, F.; JUNG, W. J.; CHARLES, T.; CLARK, S. R.; et al. The class IId bacteriocin thuricin-17 increases plant growth. **Plant**, v. 229, p. 747-755, 2009b.

LEE, S. J.; PARK, S. Y.; LEE, J. J.; YUM, D. Y.; KOO, B. T.; LEE, J. K. Genes encoding the N-acyl homoserine lactone-degrading enzyme are widespread in many subspecies of *Bacillus thuringiensis*. **Appl Environ Microbiol**, v. 68, p. 39193924, 2002.

LIBRADO, P.; ROZAS, J. DnaSP v5: a software for comprehensive analysis of DNA polymorphism data. **Bioinformatics**, v. 25, p. 1451-1452, 2009.

LIU, G.; SONG, L.; SHU, C.; WANG, P.; DENG, C.; PENG, Q.; et al. Complete genome sequence of *Bacillus thuringiensis* subsp. *kurstaki* strain HD73. **Genome Announc**, v. 1, p. e0008013, 2013.

LIU, M.; CAI, Q. X.; LIU, H. Z.; ZHANG, B. H.; YAN, J. P.; YUAN, Z. M. Chitinolytic activities in *Bacillus thuringiensis* and their synergistic effects on larvicidal activity. **J Appl Microbiol**, v. 93, p. 374-379, 2002.

LIU, X.; ZHOU, R.; FU, G.; ZHANG, W.; MIN, Y.; TIAN, Y.; et al. Draft genome sequence of *Bacillus thuringiensis* NBIN-866 with high nematocidal activity. **Genome Announc**, v. 2, p.1-2, 2014.

LOPES, R. B. M; COSTA, LEO; VANETTI, M. C. D; ARAUJO E. F.; QUEIROZ M. V. Endophytic bacteria isolated from common bean *(Phaseolus vulgaris)* exhibiting high variability showed antimicrobial activity and quorum sensing inhibition. **Curr Microbiol**, v. 71, p. 509-516, 2015.

LOPEZ-LOPEZ, A.; ROGEL, M. A.; ORMENO-ORRILLO, E.; MARTiNEZ- ROMERO, J.; MARTiNEZ-ROMERO, E. *Phaseolus vulgaris* seed-borne endophytic community with novel bacterial species such as *Rhizobium endophyticum* sp. nov. **Syst Appl Microbiol**, v. 33, p. 322-327, 2010.

LOWE, T. M.; EDDY, S. R. tRNAscan-SE: a program for improved detection of transfer RNA genes in genomic sequence. **Nucleic Acids Res**, v. 25, p. 955-964, 1997.

LUKASHIN, A. V.; BORODOVSKY, M. GeneMark.hmm: new solutions for gene finding. **Nucleic Acids Res**, v. 26, p. 1107-1115, 1998.

LUO, C.; HU, G.; ZHU, H. Genome reannotation of *Escherichia coli* CFT073 with new insights into virulence. **BMC Genomics**, v. 10, p. 552, 2009.

MACLEAN, D.; JONES, J. D.; STUDHOLME, D. J. Application of 'nextgeneration' sequencing technologies to microbial genetics. **Nat Rev Microbiol**, v. 7, p. 287-296, 2009.

MADUPU, R.; BRINKAC, L. M.; HARROW, J.; WILMING, L. G.;

BOHME, U.; LAMESCH, P.; et al. Meeting report: a workshop on best practices in genome annotation. **Database (Oxford)**, v. 2010, p. baq001, 2010.

MARCHLER-BAUER, A.; LU, S.; ANDERSON, J. B.; CHITSAZ, F.; DERBYSHIRE, M. K.; DEWEESE-SCOTT, C.; et al. CDD: a Conserved Domain Database for the functional annotation of proteins. **Nucleic Acids Res**, v. 39, p. D225-229, 2011.

MARCO, G.; MANUEL, P. Ecological mysteries: is *Bacillus thuringiensis* a real insect pathogen? **Bt Res.**, v.3, p. 1-2, 2012.

MARCON, J.; TAKETANI, R. G.; DINI-ANDREOTE, F.; MAZZERO, G. I.; SOARES, F. L. J.; MELO, I. S.; et al. Draft genome sequence of *Bacillus thuringiensis* strain BrMgv02-JM63, a chitinolytic bacterium isolated from oil- contaminated mangrove soil in Brazil. **Genome Announc**, v. 2, p. 2014.

MARDIS, E. R. The impact of next-generation sequencing technology on genetics. **Trends Genet**, v. 24, p. 133-141, 2008.

MARTIN, R. G.; ROSNER, J. L. The AraC transcriptional activators. **Curr Opin Microbiol**, v. 4, p. 132-137, 2001.

MARTÍNEZ, L.; CABALLERO-MELLADO, J.; OROZCO, J.; MARTÍNEZ- ROMERO, E. Diazotrophic bacteria associated with banana *(Musa* spp.). **Plant Soil**, v. 257, p. 35-47, 2003.

*MARTÍNEZ-ROMERO*, E. Coevolution in *Rhizobium-legume* symbiosis? **DNA Cell Biol.**, v. 28, p. 361-370, 2009.

MARTÍNEZ-ROMERO, E. Diversity of *Rhizobium-Phaseolus vulgaris* symbiosis: overview and perspectives. **Plant Soil**, v. 252, p. 11-23, 2003.

MCINROY, J. A.; KLOEPPER, J. W. Population dynamics of endophytic bacteria in field-grown sweet corn and cotton. **Can J Microbiol**, v. 41, p. 895-901, 1995.

MEADOWS, M. P. *Bacillus thuringiensis* in the environment: ecology and risk assessment. In: ENTWISTLE, P. F.; CORY, J. S.; BAILEY, M. J.; HIGGS, S. (ed). ***Bacillus thuringiensis* an environment biopesticide: theory and practice**, Chichester: John Wiley & Sons, p. 193-220, 1993.

MEDEMA, M. H.; BLIN, K.; CIMERMANCIC, P.; DE JAGER, V.; ZAKRZEWSKI, P.; FISCHBACH, M. A.; et al. Antismash: Rapid identification, annotation and analysis of secondary metabolite biosynthesis gene clusters in bacterial and fungal genome sequences. **Nucleic Acids Res**, v. 39, p.339-346, 2011.

MÉDIGUE, C.; MOSZER, I. Annotation, comparison and databases for hundreds of bacterial genomes. **Res Microbiol**, v. 158, p. 724-736, 2007.

MEDINI, D.; DONATI, C.; TETTELIN, H.; MASIGNANI, V.; RAPPUOLI, R. The microbial pan-genome. **Curr Opin Genet Dev**, v. 15, p. 589-594, 2005.

MEURER, G.; GERLITZ, M.; WENDT-PIENKOWSKI, E.; VINING, L. C.; ROHR, J.; HUTCHINSON, C. R. Iterative type II polyketide synthases, cyclases and ketoreductases exhibit context-dependent behavior in the biosynthesis of linear and angular decapolyketides. **Chem Biol**, v. 4, p. 433-443, 1997.

MISAGHI, I. J.; DONNDELINGER, C. R. Endophytic bacteria in symptomfree cottton plants. **Phytopathology**, v. 80, p. 808-811, 1990.

MISHRA, P. K.; MISHRA, S.; SELVAKUMAR, G.; BISHT, J. K.; KUNDU, S.; GUPTA, H.S. Coinoculation of *Bacillus thuringiensis-KR1* with *Rhizobium leguminosarum* enhances plant growth and nodulation of pea *(Pisum sativum* L.) and lentil *(Lens culinaris* L.). **World J Microbiol Biotechnol**, v. 25, p. 753-761, 2009.

MISHRA, R.; SINGH, R.; JAISWAL, H.; KUMAR, V.; MAURYA, S. *Rhizobium-mediated* induction of phenolics and plant growth promotion in rice (*Oryza sativa* L.). **Curr Microbiol**, v. 52, p. 383-389, 2006.

MONNERAT, R. G.; SANTOS, R.; BARROS, P.; BATISTA, A.; BERRY, C. **Isolation and characterization of endophytic *Bacillus thuringiensis* strains of cotton,** Brasília: Embrapa Genetic Resources and Biotechnology, 4 p., 2003.

MURAWSKA, E.; FIEDORUK, K.; BIDESHI, D. K.; SWIECICKA, I. Complete genome sequence of *Bacillus thuringiensis* subsp. *thuringiensis* strain IS5056, an isolate highly toxic to *Trichoplusia ni*. **Genome Announc**, v. 1, p. e0010813, 2013.

NAVON, A. *Bacillus thuringiensis* application in agriculture. In: CHARLES, J. F.; DELÉCLUSE, A.; NILESEN-LE ROUX, C. (ed.).

**Entomopathogenic bacteria: from laboratory to field application**, Netherlands: Kluwer academic Publishers, p. 355-367, 2000.

OVERBEEK, R.; BEGLEY, T.; BUTLER, R. M.; CHOUDHURI, J. V.; CHUANG, H. Y.; COHOON, M.; et al. The subsystems approach to genome annotation and its use in the project to annotate 1000 genomes. **Nucleic Acids Res**, v. 33, p. 5691-5702, 2005.

PAIK, H. D.; BAE, S. S.; PARK, S. H.; PAN, J. G. Identification and partial characterization of tochicin, a bacteriocin offduced by *Bacillus thuringiensis* subsp. *tochigiensis*. **J Ind Microbiol Biotechnol**, v. 19, p. 294-298, 1997.

PALMA, L.; MUNOZ, D.; BERRY, C.; MURILLO, J.; CABALLERO, P. Draft genome sequences of two *Bacillus thuringiensis* strains and characterization of a putative 41.9-kDa insecticidal toxin. **Toxins (Basel)**, v. 6, p. 1490-1504, 2014.

PARK, R. Y.; CHOI, M. H.; SUN, H. Y.; SHIN, S. H. Production of catecholsiderophore and utilization of transferrin-bound iron in *Bacillus cereus*. **Biol Pharm Bull**, v. 28, p. 1132-1135, 2005.

PATEL, P. S.; HUANG, S.; FISHER, S.; PIRNIK, D.; AKLONIS, C.; DEAN,
L. ; et al. Bacillaene, a novel inhibitor of prokaryotic protein synthesis produced by Bacillus subtilis: production, taxonomy, isolation, physico-chemical characterization and biological activity. **J Antibiot (Tokyo)**, v. 48, p. 997-1003, 1995.

PAULSEN, I. T.; PRESS, C. M.; RAVEL, J.; KOBAYASHI, D. Y.; MYERS, G. S.; MAVRODI, D. V.; et al. Complete genome sequence of the plant commensal *Pseudomonas fluorescens* Pf-5. **Nat Biotechnol**, v. 23, p. 873-878, 2005.

POLANCZYK, R; ALVES, S. *Bacillus thuringiensis:* a brief review. **Agrociência**, v. 8, p. 1-10, 2003.

PRAÇA, L. B.; GOMES, A. C. M. M.; CABRAL, G.; MARTINS, E. S.; SUJII, E. R.; MONNERAT, R. G. Endophytic colonization by brazilian strains of *Bacillus thuringiensis* on cabbage seedlings grown *in vitro*. **Bt Res**, v. 3, p. 11-19, 2012.

PRICE, A. L.; JONES, N. C.; PEVZNER, P. A. De novo identification of repeat families in large genomes. **Bioinformatics**, v. 21, p.

i351-358, 2005.

PRIEST, F. G.; BARKER, M.; BAILLIE, L. W.; HOLMES, E. C.; MAIDEN,
M . C. Population structure and evolution of the *Bacillus cereus* group. **J Bacteriol**, v. 186, p. 7959-7970, 2004.

QUADT-HALLMANN, A.; KLOEPPER, J. W. Immunological detection and localization of the cotton endophyte *Enterobacter asburiae* JM22 in different plant species. **Can J Microbiol**, v. 42, p. 1144-1154, 1996.

QUIÑONES, B.; DULLA, G.; LINDOW, S. E. Quorum sensing regulates exopolysaccharide production, motility and virulence in *Pseudomonas syringae*. **Mol Plant Microbe Interact**, v 18, p. 682-693, 2005.

RADDADI, N; CHERIF, A.; OUZARI, H.; MARZORATI, M.; BRUSETTI, L.; et al. *Bacillus thuringiensis* beyond insect biocontrol: plant growth promotion and biosafety of polyvalent strains, **Ann Microbiol**, v. 57, p. 481-494, 2007.

RASKO, D. A.; ROSOVITZ, M. J.; MYERS, G. S.; MONGODIN, E. F.; FRICKE, W. F.; GAJER, P.; et al. The pangenome structure of *Escherichia coli*: comparative genomic analysis of *E. coli* commensal and pathogenic isolates. **J Bacteriol**, v. 190, p. 6881-6893, 2008.

REA, M. C.; ALEMAYEHU, D.; CASEY, P. G.; O'CONNOR, P. M.; LAWLOR P. G.; WALSH. M.; et al. Bioavailability of the anti-clostridial bacteriocin thuricin CD in gastrointestinal tract.**Microbiology**, v. 160, p. 439-445, 2014.

READ, T. D.; AKMAL, A.; BISHOP-LILLY, K. **Annotation of the *Bacillus thuringiensis* BGSC4Y1 genome**. Available at: http://www.ncbi.nlm.nih.gov/nuccore/NZ CM000746.1, 2009.

REESE, M. G. Application of a time-delay neural network to promoter annotation in the *Drosophila melanogaster* genome. **Comput Chem**, v. 26, p. 51-56, 2001.

REEVES, G. A.; TALAVERA, D.; THORNTON, J. M. Genome and proteome annotation: organization, interpretation and integration. **J R Soc Interface**, v. 6, p. 129-147, 2009.

RISSMAN, A. I.; MAU, B.; BIEHL, B. S.; DARLING, A. E.; GLASNER, J. D.; PERNA, N. T. Reordering contigs of draft genomes using the Mauve aligner. **Bioinformatics**, v. 25, p. 2071-2073, 2009.

ROH, J. Y.; CHOI, J. Y.; LI, M. S.; JIN, B. R.; JE, Y. H. Bacillus thuringiensis as a specific, safe, and effective tool for insect pest control. **J Microbiol Biotechnol**, v. 17, p. 547-559, 2007.

ROJAS-AVELIZAPA, L. I.; CRUZ-CAMARILLO, R.; GUERRERO, M. I.; RODRIGUEZ-VAZQUEZ, R.; IBARRA, J. E. Selection and characterization of a proteo-chitinolytic strain of *Bacillus thuringiensis,* able to grow in waste media. **World J Microbiol Biotechnol**, v. 15, p. 261-268, 1999.

ROULI, L.; M, M. B.; ROBERT, C.; NDIAYE, M.; LA SCOLA, B.; RAOULT, D. Genomic analysis of three African strains of *Bacillus anthracis* demonstrates that they are part of the clonal expansion of an exclusively pathogenic bacterium. **New Microbes New Infect**, v. 2, p. 161-169, 2014.

ROSEN BLUETH, M.; MARTINEZ-ROMERO, E. Bacterial endophytes and their interactions with hosts. **Mol Plant Microbe Interact**, v. 19, p. 827-837, 2006.

RUDE, M. A; KHOSLA, C. Engineered biosynthesis of polyketides in heterologous hosts. **Chem Eng Sci**, v. 59, p. 4693-4701, 2004.

RUST, A. G.; MONGIN, E.; BIRNEY, E. Genome annotation techniques: new approaches and challenges. **Drug Discov Today**, v. 7, p. S70-76, 2002.

RUTHERFORD, K.; PARKHILL, J.; CROOK, J.; HORSNELL, T.; RICE, P.; RAJANDREAM, M. A.; et al. Artemis: sequence visualization and annotation. **Bioinformatics**, v. 16, p. 944-945, 2000.

RYAN, R. P.; GERMAINE, K.; FRANKS, A.; RYAN, D. J.; DOWLING, D.
N . Bacterial endophytes: recent developments and applications. **FEMS Microbiol Lett**, v. 278, p. 1-9, 2008.

SADFI, N.; CHERIF, M.; FLISS, I.; BOUDABBOUS, A.; ANTOUN, H. Evaluation of *Bacillus* isolates from salty soils and *Bacillus thuringiensis* strains for the biocontrol of *Fusarium* dry rot of potato tubers. **J Plant Pathol**, v. 83, p. 101-118, 2001.

SALAMA, H. S.; MORRIS, O. N. The use of *Bacillus thuringiensis* in development countries. In: ENTWISTLE, P. F.; CORY, J. S.; BAILEY, M. J.; HIGGS, S. (ed). ***Bacillus thuringiensis,* an environmental biopesticide: theory and practice**, Chichester: John Wiley & Sons, p. 237-253, 1993.

SALZBERG, S. L. Genome re-annotation: a wiki solution? **Genome Biol**, v. 8, p. 102, 2007.

SANGER, F.; AIR, G. M.; BARRELL, B. G.; BROWN, N. L.; COULSON, A. R.; FIDDES, C. A.; et al. Nucleotide sequence of bacteriophage phi X174 DNA. **Nature**, v. 265, p. 687-695, 1977a.

SANGER, F.; NICKLEN, S.; COULSON, A. R. DNA sequencing with chainterminating inhibitors. **Proc Natl Acad Sci USA**, v. 74, p. 5463-5467, 1977b.

SCHNEPF, E.; CRICKMORE, N.; VAN RIE, J.; LERECLUS, D.; BAUM, J.; FEITELSON, J.; et al. *Bacillus thuringiensis* and its pesticidal crystal proteins. **Microbiol Mol Biol Rev**, v. 62, p. 775-806, 1998.

SCOTT, B. Epichloe endophytes: fungal symbionts of grasses. **Curr Opinin Microbiol**, v. 4, p. 393-398, 2001.

SHEPPARD, A. E.; POEHLEIN, A.; ROSENSTIEL, P.; LIESEGANG, H.; SCHULENBURG, H. Complete genome sequence of *Bacillus thuringiensis* strain 407 *cry-*. **Genome Announc**, v. 1, p. 1-2, 2013.

SMITH, R. A; COUCHE, G. A. The phylloplane as a source of *Bacillus thuringiensis* variants. **Appl Environ Microbiol**, v. 57, p. 311-315, 1991.

SMITH, R. A; COUCHE, G. A. The phylloplane as a source of *Bacillus thuringiensis* variants. **Appl Environ Microbiol**, v. 57, p. 311-315, 1991.

SMITS, T. H.; JAENICKE, S.; REZZONICO, F.; KAMBER, T.; GOESMANN, A.; FREY, J. E.; et al. Complete genome sequence of the fire blight pathogen *Erwinia pyrifoliae* DSM 12163T and comparative genomic insights into plant pathogenicity. **BMC Genomics**, v. 11, p. 2, 2010.

SNEH, B.; SCHUSTER S.; GROSS, S. Improvement of the insecticidal activity of *Bacillus thuringiensis* var. *entomocidus* on larvae of *Spodoptera fittoralis* (Lepidoptera, Noctuidae) by addition of chitinolytic bacteria, a phagostimulant and a UV-protectant. **Z Angew Entomol**, v. 96,

p. 77-83, 1983.

SPRENT, J.; FARIA, S. Mechanisms of infection of plants by nitrogen fixing organisms. **Plant Soil**, v. 110, p. 157-165, 1988.

STAHLY, P.; ANDREWS, R. E.; YOUSTEN, A. A. The genus *Bacillus* - Insect Pathogens. In: BALOWS, A. H. G.; TRUPPER, M.; DWORKIN, W; HERDER, K.H. **The Prokaryotes,** 2$^{nd}$ ed., New York: Springer-Verlag, p. 4770, 1991.

STRAGIER, P.; KUNKEL, B.; KROOS, L.; LOSICK, R. Chromosomal rearrangement generating a composite gene for a developmental transcription factor. **Science**, v. 243, p. 507-512, 1989.

STEIN, L. Genome annotation from sequence to biology. **Nat Rev**, v. 2, p. 493-505, 2001.

STUDHOLME, D. J.; DIXON, R. Domain architectures of sigma54-dependent transcriptional activators. **J Bacteriol**, v. 185, p. 1757-1767, 2003.

SUBRAHMANYAN, P.; REDDY, M. N.; RAO, A. S. Exudation of certain organic compounds from seeds of groundnut. **Seed Sci Technol**, v. 11, p. 267-27, 1983.

SUGINTA, W.; ROBERTSON, P. A.; AUSTIN, B.; FRY, S. C.; FOTHERGILL-GILMORE, L. A. Chitinases from Vibrio: activity screening and purification of *chiA* from *Vibrio carchariae*. **J Appl Microbiol**, v. 89, p. 76-84, 2000.

SULLIVAN, T. J.; RODSTROM, J.; VANDOP, J.; LIBRIZZI, J.; GRAHAM, C.; SCHARDL, C. L., et al. Symbiont-mediated changes in *Lolium arundinaceum* inducible defenses: evidence from changes in gene expression and leaf composition. **New Phytol**, v. 176, p. 673-679, 2007.

SUN, Y.; ZHAO, Q.; XIA, L.; DING, X.; HU, Q.; FEDERICI, B. A.; et al. Identification and characterization of three previously undescribed crystal proteins from *Bacillus thuringiensis* subsp. *jegathesan*. **Appl Environ Microbiol**, v. 79, p. 3364-3370, 2013.

SUZUKI, M. T.; HERNÁNDEZ-RODRÍGUEZ, C. S.; ARAÚJO, W. L.; FERRÉ, J. Characterization of an endophytic *Bacillus thuringiensis* strain isolated from sugar cane. In: **Proceedings of 41st Annual Meeting of the Society for Invertebrate Pathology and 9th International**

**Conference on** ***Bacillus thuringiensis,*** Coventry, United Kingdom, 3-7 August 2008.

SUZUKI, M. T.; LERECLUS, D.; ARANTES, O. M. N. Fate of *Bacillus thuringiensis* strains in different insect larvae. **Can J Microbiol**, v. 50, p. 973-975, 2004.

TAMURA, K.; STECHER, G.; PETERSON, D.; FILIPSKI, A.; KUMAR, S. MEGA6: Molecular Evolutionary Genetics Analysis version 6.0. **Mol Biol Evol**, v. 30, p. 2725-2729, 2013.

TAN, F.; ZHENG, A.; ZHU, J.; WANG, L.; LI, S.; DENG, Q.; WANG, S.; LI, P.; TANG, X. Rapid cloning, identification, and application of one novel crystal protein gene *cry30Fa1* from *Bacillus thuringiensis.* **FEMS Microbiol Lett**, v. 302, p. 46-51, 2010.

TAN, F.; ZHU, J.; TANG, J.; TANG, X.; WANG, S.; ZHENG, A.; LI, P. Cloning and characterization of two novel crystal protein genes, *cry54Aa1* and *cry30Fa1,* from *Bacillus thuringiensis* strain BtMC28. **Curr Microbiol**, v. 58, p. 654-659, 2009.

TANG, Y.; ZOU, J.; ZHANG, L.; LI, Z.; MA, C.; MA, N. Anti-fungi activities of *Bacillus thuringiensis* H3 chitinase and immobilized chitinase particles and their effects to rice seedling defensive enzymes. **J Nanosci Nanotechnol**, v. 12, p. 80818086, 2012.

TANUJA, R.; BISHT, S. C.; MISHRA, P. K. Ascending migration of endophytic *Bacillus thuringiensis* and assessment of benefits to different legumes of N.W. Himalayas. **Eur J Soil Biol**, v. 56, p. 56-64, 2012.

TETTELIN, H.; MASIGNANI, V.; CIESLEWICZ, M. J.; DONATI, C.; MEDINI, D.; WARD, N. L.; et al. Genome analysis of multiple pathogenic isolates of *Streptococcus agalactiae:* implications for the microbial "pan-genome". **Proc Natl Acad Sci USA**, v. 102, p. 13950-13955, 2005.

THOMAS, D. J. I.; ALUN, J.; MORGAN, W.; WHIPPS, J. M.; SAUNDERS, J. R. Plasmid transfer betweem the *Bacillus thuringiensis* subspecies *kurstaki* and *tenebrionis* in laboratory culture and soil and in Lepidopteran and Coleopteran larvae. **Appl Environ Microbiol**, v. 66, p. 118-124, 2000.

TICKNOR, L. O.; KOLSTO, A. B.; HILL, K. K.; KEIM, P.; LAKER, M. T.; TONKS, M.; et al. Fluorescent amplified fragment length polymorphism analysis of Norwegian *Bacillus cereus* and *Bacillus thuringiensis* soil isolates. **Appl Environ Microbiol**, v. 67, p. 4863-4873, 2001.

TILLIER, E. R.; COLLINS, R. A. Genome rearrangement by replication- directed translocation. **Nat Genet**, v. 26, p. 195-197, 2000.

TURNBAUGH, P. J.; LEY, R. E.; HAMADY, M.; FRASER-LIGGETT, C. M.; KNIGHT, R.; GORDON, J. I. The human microbiome project. **Nature**, v. 449, p. 804-810, 2007.

VAN FRANKENHUZEN, K. The challenge of *Bacillus thuringiensis*. In: ENTWISTLE, P. F.; CORY, J. S.; BAILEY, M. J.; HIGGS, S. (ed). ***Bacillus thuringiensis*, an environment biotpesticide: theory and practice**, Chichester: John Wiley & Sons, p. 1-35. 1993.

VAN HEEL, A. J; DE JONG, A.; MONTALBAN-LOPEZ, M.; KOK, J.; KUIPERS, O. P. BAGEL3: Automated identification of genes encoding bacteriocins and (non-) bactericidal posttranslationally modified peptides. **Nucleic Acids Res**, v. 41, p. 448-453, 2013.

VAN LANEN, S. G.; SHEN, B. Microbial genomics for the improvement of natural product discovery. **Curr Opin Microbiol**, v. 9, p. 252-260, 2006.

VENTER, J. C.; ADAMS, M. D.; MYERS, E. W.; LI, P. W.; MURAL, R. J.; SUTTON, G. G.; et al.. The sequence of the human genome. **Science**, v. 291, p. 1304-1351, 2001.

VILAS-BOAS, G. F. L. T.; VILAS-BOAS, L. A.; LERECLUS, D.; ARANTES, O. M. N. *Bacillus thuringiensis* conjugation under environmental conditions. **FEMS Microbiol Ecol**, v. 25, p. 369-374, 1998.

VILAS-BOAS, L. A.; VILAS-BOAS, G. F. L. T.; SARIDAKIS, H. O.; LEMOS, M. V. F.; LERECLUS, D.; ARANTES, O. M. N. Survival and conjugation of *Bacillus thuringiensis* in a soil microcosm. **FEMS Microbiol Ecol**, v. 31, p. 225259, 2000.

VON BODMAN, S. B.; BAUER, W. D.; COPLIN, D. L. Quorum sensing in plant-pathogenic bacteria. **Annu Rev Phytopathol**, v. 41, p. 455-482, 2003.

WANG, A.; PATTEMORE, J.; ASH, G.; WILLIAMS, A.; HANE, J. Draft genome sequence of *Bacillus thuringiensis* strain DAR 81934, which exhibits molluscicidal activity. **Genome Announc**, v. 1, p. 1-2, 2013.

WANG, S. L.; HWANG, J. Microbial reclamation of shellfish wastes for the production of chitinases. **Enzyme Microb Technol**, v. 28, p. 376-382, 2001.

WANG, S. L.; LIN, T. Y.; YEN, Y. H.; LIAO, H. F.; CHEN, Y. J. Bioconversion of shellfish chitin wastes for the production of *Bacillus subtilis* W-118 chitinase. **Carbohydr Res**, v. 341, p. 2507-2515, 2006.

WEI, J. Z.; HALE, K.; CARTA, L.; PLATZER, E.; WONG, C.; FANG, S. C.; et al. Bacillus thuringiensis crystal proteins that target nematodes. **Proc Natl Acad Sci U S A,** v. 100, p. 2760-2765, 2003.

WEISER, J. Impact of *Bacillus thuringiensis* on applied entomology in eastern Europe and in Soviet Union. In: KRIEG, A.; HUGER, A. **M. Mitteilungen aus der biologischen bundesanstalt für land und forstwirtschaft Berlin-Dahlem heft**, Berlin: Paul Parey, p. 37-50, 1986.

WILSON, K. E.; FLOR, J. E.; SCHWARTZ, R. E.; JOSHUA, H.; SMITH, J. L.; PELAK, B. A.; et al. Difficidin and oxydifficidin: novel broad spectrum antibacterial antibiotics produced by *Bacillus subtilis*. II. Isolation and physicochemical characterization. **J Antibiot (Tokyo)**, v. 40, p. 1682-1691, 1987.

WILSON, M. K.; ABERGEL, R. J.; RAYMOND, K. N.; ARCENEAUX, J. E.; BYERS, B. R. Siderophores of *Bacillus anthracis, Bacillus cereus,* and *Bacillus thuringiensis*. **Biochem Biophys Res Commun**, v. 348, p. 320-325, 2006.

WIWAT, C.; THAITHANUN, S.; PANTUWATANA, S.; BHUMIRATANA, A. Toxicity of chitinase-producing Bacillus *thuringiensis* ssp. *kurstaki* HD-1 (G) toward *Plutella xylostella*. **J Invertebr Pathol**, v. 76, p. 270-277, 2000.

YE, W.; ZHU, L.; LIU, Y.; CRICKMORE, N.; PENG, D.; RUAN, L.; et al. Mining new crystal protein genes from *Bacillus thuringiensis* on the basis of mixed plasmid-enriched genome sequencing and a computational pipeline. **Appl Environ Microbiol**, v. 78, p. 4795-4801, 2012.

ZAWADZKA, A. M.; KIM, Y.; MALTSEVA, N.; NICHIPORUK, R.; FAN, Y.; JOACHIMIAK, A.; et al. Characterization of a *Bacillus subtilis* transporter for petrobactin, an anthrax stealth siderophore. **Proc Natl Acad Sci USA**, v. 106, p. 21854-21859, 2009.

ZGHAL, R. Z.; JAOUA, S. Evidence of DNA rearrangements in the 128kilobase pBtoxis plasmid of *Bacillus thuringiensis israelensis*. **Mol Biotechnol**, v. 33, p. 191-198, 2006.

ZHANG, G.; DENG, A.; XU, Q.; LIANG, Y.; CHEN, N.; WEN, T. Complete genome sequence of *Bacillus amyloliquefaciens* TA208, a strain for industrial production of guanosine and ribavirin. **J Bacteriol**, v. 193, p. 3142-3143, 2011.

ZHANG, W.; CHEN, J.; YANG, Y.; TANG, Y.; SHANG, J.; SHEN, B. A practical comparison of de novo genome assembly software tools for next-generation sequencing technologies. **Plos One**, v. 6, p. e17915, 2011.

ZHOU, Y.; CHOI, Y. L.; SUN, M.; YU, Z. Novel roles of Bacillus *thuringiensis* to control plant diseases. **Appl Microbiol Biotechnol**, v. 80, p. 563572, 2008.

ZHU, Y.; SHANG, H.; ZHU, Q.; JI, F.; WANG, P.; FU, J.; et al. Complete genome sequence of *Bacillus thuringiensis* serovar *finitimus* strain YBT-020. **J Bacteriol**, v. 193, p. 2379-2380, 2011.

ZIMMERMAN, S. B.; SCHWARTZ, C. D.; MONAGHAN, R. L.; PELAK, B. A.; WEISSBERGER, B.; GILFILLAN, E. C; et al. Difficidin and oxydifficidin: novel broad spectrum antibacterial antibiotics produced by *Bacillus subtilis*. I. Production, taxonomy and antibacterial activity. **J Antibiot (Tokyo)**, v. 40, p. 16771681, 1987.

ZWEERINK, M. M.; EDISON, A. Difficidin and oxydifficidin: novel broad spectrum antibacterial antibiotics produced by *Bacillus subtilis*. III. Mode of action of difficidin. **J Antibiot (Tokyo)**, v. 40, p. 1692-1697, 1987.

ZWICK, M. E.; JOSEPH, S. J.; DIDELOT, X.; CHEN, P. E.; BISHOP-LILLY, K. A.; STEWART, A. C.; et al. Genomic characterization of the *Bacillus cereus sensu lato* species: backdrop to the evolution of *Bacillus anthracis*. **Genome Res**, v. 22, p. 1512-1524, 2012.

MIX
Papier aus verantwortungsvollen Quellen
Paper from responsible sources
FSC® C105338
FSC
www.fsc.org